Proceedings in Life Sciences

Toxins, Drugs, and Pollutants in Marine Animals

Edited by

L. Bolis J. Zadunaisky R. Gilles

With 67 Figures

Springer-Verlag
Berlin Heidelberg New York Tokyo 1984

Professor Dr. L. Bolis (Scientific Editor Part B)
Institute of General Physiology
University of Messina, Via dei Verdi 85
I-98100 Messina, Italy

Professor Dr. J. Zadunaisky (Scientific Editor Part A)
Department of Physiology and Biophysics
New York University Medical Center
550 First Avenue
New York, NY 10016, USA

Professor Dr. R. Gilles (Coordinating Editor)
Laboratory of Animal Physiology
University of Liège, 22, Quai van Beneden
B-4020 Liège, Belgium

For explanation of the cover motive see Scheme 1, page 6.

ISBN-13:978-3-642-69905-4 e-ISBN-13:978-3-642-69903-0
DOI: 10.1007/978-3-642-69903-0

Library of Congress Cataloging in Publication Data. Main entry under title: Toxins, drugs, and pollutants in marine animals. (Proceedings in life sciences) Includes index. 1. Marine fauna – Effect of water pollution on. 2. Marine toxins. 3. Marine pharmacology. I. Bolis, Liana. II. Zadunaisky, José A. III. Gilles, R. IV. Series. QL121.T69 1984 591.5'2636 84-16054

Softcover reprint of the hardcover 1st edition 1984

2131/3130-543210

Foreword

The importance of toxins and other pharmacologically active compounds obtained from marine animals cannot be emphasized enough. The use of these chemicals for defense or attack of other life forms are of interest in themselves; however, they are of great importance in our understanding of their mechanisms of action in view of possible application in the defense of man (no doubt a predator) and in biology and medicine. The toxin of the flat fish *Pardachirus* presented in some of the papers of this book is an example of a defense mechanism based on secretion of a toxin that could possibly be extended to being used to defend man himself from sharks, the marine predators par excellence.

August, 1984 J. ZADUNAISKY

Preface

The study of toxins, drugs, and pollutants in the marine environment, and their impact on human life existed already in Babylon and Assyria, Egypt, Persia, India, China, Japan, Greece, Rome and South America. Nevertheless, less is known about ethnic marine biology than about ethnobotany. Only recently however, have active biotoxins been used as molecular probes in neuropharmacology, offering intriguing new insights into nervous activity and muscular functions.

It is clear from the presentations at this meeting that much remains to be done, and certainly new, more pharmacologically oriented findings lie ahead.

Toxic effects of pollutants are being studied increasingly especially for their involvement in respiratory and osmoregulatory systems at fish gill level and in invertebrates. The importance of solubility of pollutants in seawater should be taken into account, since the physico-chemical features of the marine environment are by no means uniform.

Another particular aspect of pollution is acid rain, the name given to the wet precipitation of pollutants SO2/SO3 and NO2/HNO3. Although this kind of pollution is more relevant to freshwater, woods, and buildings, it nevertheless also affects the sea environment.

The main effect of acid rain on teleosteons and invertebrates appears to be the alteration of the membrane composition at the gill level where the main changes induced are in the plasmalogens; at present, however, useful studies can only be performed on freshwater teleosteons.

The Symposium was also supplemented by shorter presentations and poster sessions centered on the central theme.

August, 1984 L. BOLIS

Preface

Contents

Part A Toxins and Drugs of Marine Animals

Part B Responses of Marine Animals to Pollutants

List of Contributors

You will find the addresses at the beginning of the respective contribution

Part A
Toxins and Drugs of Marine Animals

Shark-Repelling Ichthyotoxins from the Defensive Secretion of the Sole *Pardachirus pavoninus*

K. TACHIBANA[1]

1 Introduction

Toxic fish are categorized into three groups by Halstead (Halstead 1978). The first group, termed poisonous fish, comprises those causing fish poisoning when eaten by humans, and is classified according to location of the toxin. Common fish poisoning, such as puffer poisoning and ciguatera, is caused by eating the flesh, and the fish responsible are called ichthyosarcotoxic; when gonad or blood cause poisoning, the fish in these cases are called ichthyotoxic and ichthyohemotoxic respectively. In contrast to poisonous fish, which have no specialized poison gland and whose toxins, most probably formed exogenously, are not positively used in self-defense, the second group venomous fish, also called acanthotoxic fish, have poison glands associated with traumagenic devices, usually highly developed pungent fin spines. Their toxins are exclusively proteinaceous as far as known to date, whereas the toxins causing fish poisoning are ussually not (Hashimoto 1979).

Some other fish have special secretory cells which produce ichthyotoxic substances, in addition to common mucous cells universally found among the class Pisces, but lack any traumagenic devices. These fish constitute the third group, termed ichthyocrinotoxic fish. They usually live in tropic reef regions where the struggle for survival is harsh, swim rather slowly or inefficiently, and lack solid squamation. On the basis of these common characteristics, the ichthyotoxic substances, which are termed ichthyocrinotoxins and also called fish mucous toxins or fish skin toxins, are thought to play a role in self-defense in the host fishes. Ichthyocrinotoxic fish have thus been suspected to be a primitive form of venomous fish, i.e., the toxic mucous cells gathering around particular fin spines, then associating and connecting with them through grooves, to make venom-injecting systems (Cammeron and Endean 1973).

Pahutoxin from a boxfish *Ostracion lentiginosus* is the first ichthyocrinotoxin whose chemical structure has been fully characterized (Boylan and Scheuer 1967). Recently, deacetoxypahutoxin, i.e., parmitoylcholin, was isolated as another ichthyocrinotoxin from the smooth trunkfish *Lactophrys triqueter* of the same trunkfish family (Goldberg et al. 1982), but aside from these exceptions, no chemical structure

1 Suntory Institute for Bioorganic Research, Shimamoto-cho, Mishima-gun, Osaka 618, Japan

Toxins, Drugs, and Pollutants in Marine Animals
(ed. by Bolis et al.)

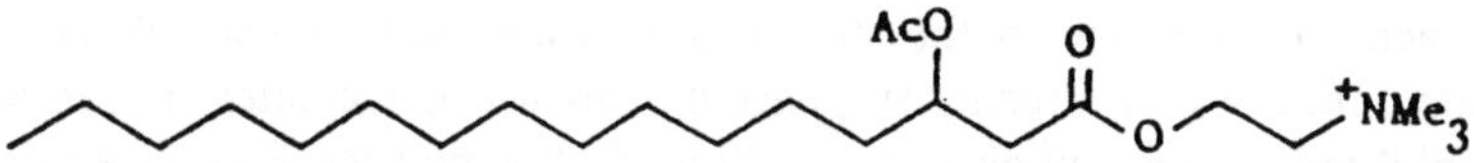

Pahutoxin

has been presented for other molecules of this category. Furthermore, all others which have been chemically investigated were either peptidic or proteinaceous. A number of these toxins belong to a group of peptides first isolated from the soapfish *Pogonoperca punctata* and named grammistins (Hashimoto and Oshima 1972). The respective mulecules are ca. 80% peptidic and each comprises a chain of 20-30 amino acids conjugated with an unknown lipophilic moiety with a quarternary amine. Similar toxins were recognized in coral-gobies (Hashimoto et al. 1974) and in the clingfish (Hori et al. 1979). A protein named pardaxin was isolated from the Red Sea Moses sole *Pardachirus marmoratus,* known as a shark-proof fish (Clark 1974), and its amino acid composition has been reported (Primor et al. 1978). Isolation of these ichthyocrinotoxins was made by monitoring ichthyotoxicity and/or hemolytic activity (Hashimoto 1979).

The sole *P. pavoninus,* a close kin of the Moses sole, dwells at the sandy bottom of tropic reefs in the western Pacific and eastern Indian Oceans, and has similar toxic cells along the base of its dorsal and anal fin spines (Clark and George 1979). Locally in Okinawa where the sole was captured for the present study, it is known to be too bitter to eat unless the skin is removed (Hashimoto 1979). The present article deals with isolation and structure elucidation of ichthyotoxic steroid monoglycosides, named pavoninins, from *P. pavoninus* secretion. In the tonic immobility test carried out at Miami, they were shown to have relatively strong shark-repelling activity.

2 Isolation of Pavoninins

Ten individuals of *P. pavoninus,* 20-30 cm long, were captured at Kabira Bay, Ishigaki Island, Ryukyu Archipelago, Japan, in July, 1981. Three of them died on the second day of captivity despite being kept in an aquarium with flowing water. From the seven left, the milky secretion was collected simply by placing the fish out of water on an enameled steel tray, where its violent effort to escape caused a secretion discharge. The fish was returned into the aquarium within a minute. The milking was performed once more from the same individuals on the next day; further milking failed, however, because the fish died after a few days as a result of air transportation in sea water, one individual each in an oxygen-inflated plastic bag, to a marine station of The Ryukyu University on the main island of Okinawa. They presumably poisoned themselves. After their death and storage-freezing, more mucus was squeezed out of their toxic cells, but the fresh secretion of the 14 discharges turned out to be sufficient for the present chemical study.

The viscous secretion collected in the steel tray was washed into a sample bottle with sea water, resulting in a foamy suspension. This was immediately put in a freezer, and lyophilized within a week to yield 30 g of white powder. Fractionation was made from this powder by monitoring ichthyotoxicity to Japanese killifish *Oryzias latipes* with a 10 ml sample solution of known concentration in 50 mM phosphate buffer (pH 7.2) or in 1% aqueous ethanol. For some samples, hemolytic activity was measured on rabbit erythrocytes by monitoring optical absorption at 560 nm of the supernatant from hemolyzed suspension. The sequence of fractionation to concentrate ichthyotoxicity is summarized in Fig. 1, where ichthyotoxicity is expressed by 1-h lethal concentration (LD_{1h}) and hemolytic activity by the relative activity to 1 mg standard saponin (E. Merck) (SU = saponin unit).

The acetone precipitate retained a little more than a half of ichthyotoxicity of the lyophilized secretion, and it appears to consist mostly of proteinaceous or peptidic substances judging from its solubilities and chromatographic behaviors. Two thirds in weight of the precipitate are dialyzable, leaving the toxicity in the remainder. In contrast to the reported instability of pardaxin (Primor and Zlotkin 1975), the active substance(s) seem rather stable, since the acitivity is tolerant through precipitation by acetone and exposure to mild acids and bases. No further study has yet been made of this hydrophilic factor.

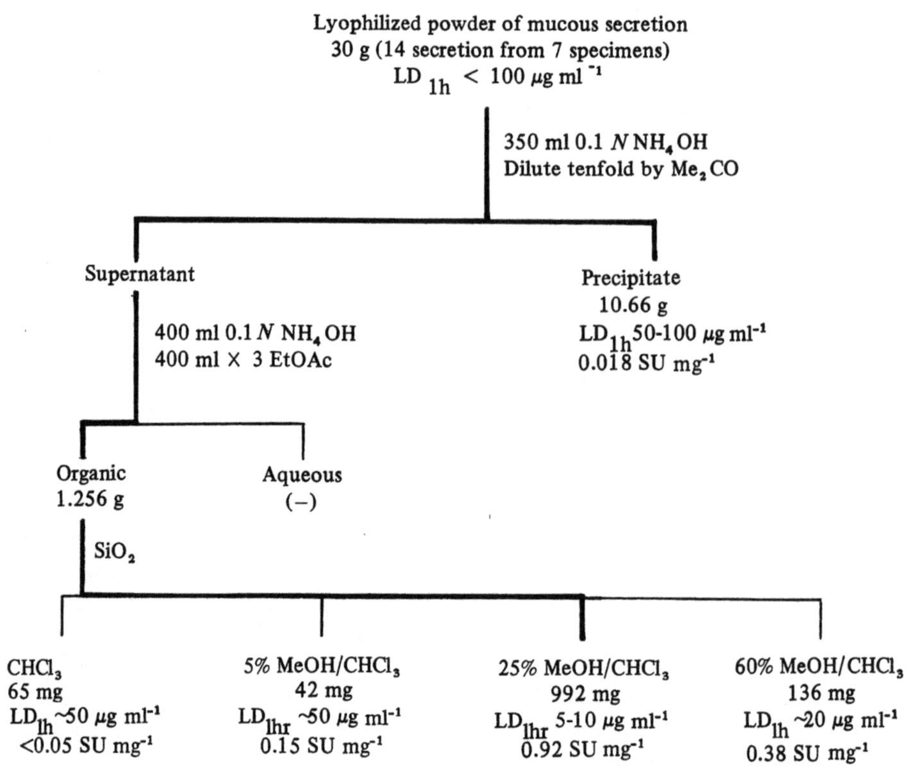

Fig. 1. Fractionation of *P. pavoninus* mucous secretion

The 25% methanol/chloroform eluate (see Fig. 1) inherited more than 90% of both the activities from the lipophilic ethyl acetate extract of the secretion, and showed two spots on silica-gel thin layer chromatography (tlc) developed by the same solvent system as the eluant at Rf 0.6 and 0.5. Its ichthyotoxicity of LD_{1h} 8.5 $\mu g\ ml^{-1}$ is not surprisingly strong, but as a single discharge contains an average of 70 mg of this material on the basis of the isolation yield, the sole can be an active fish killer. In the shark-repellency assays discussed later, "pavoninins" indicate this fraction, which was found to be a mixture of similarly toxic substances. The fraction is responsible for ca. 80% of hemolytic acitivity in the lyophilized secretion.

The mixture was first separated according to its tlc spots by silica-gel chromatography, where separation is made based on polarity of molecules in general, or the number of hydroxyls in the present case. Each of the two fractions was further separated or purified by reverse-phase (octylsilanized silica) chromatography based on lipophilicity of molecules determined by orientation of a hydroxyl and/or difference in the hydrocarbon moiety. One of the fractions thus obtained was found to be a further mixture of two compounds, and finally separated by silver nitrate-impregnated silica-gel chromatography based on different environments of the olefin

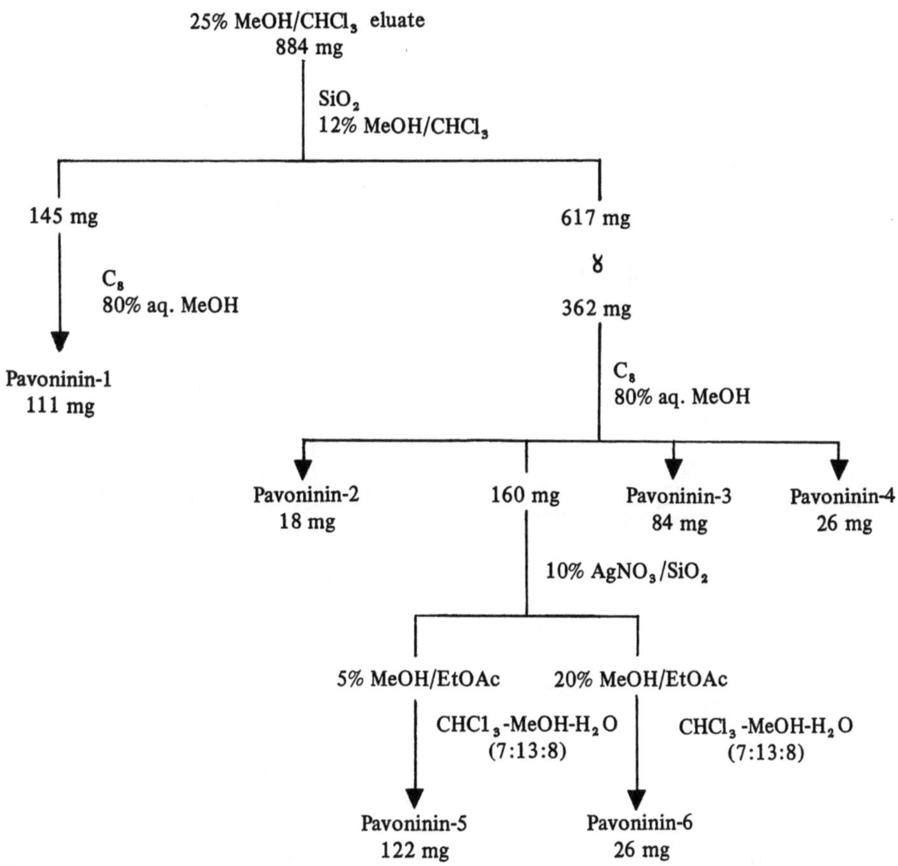

Fig. 2. Mutual separation of lipophilic toxins

in the molecules. Silver nitrate, which accompanied elution from the last column, was removed by the partition shown in Fig. 2, where mutual separation of pavoninins is summarized. The ichthyotoxicity of separated pavoninins falls within a range of LD_{1h} 5-10 μg ml^{-1}, practically equivalent values to that of the unseparated mixture.

3 Structure Elucidation of Pavoninins

The chemical structures of pavoninins were suggested to be steroid monoglycosides from their spectroscopic data, with a degree of variation in their steroidal aglycons. Chemical degradation and conversions were carried out on them to establish their full structures with entire stereochemistry to be **1-6**.

The spectroscopic data of pavoninin-1 (**1**) clarified all the functional groups in it, and paramagnetic shifts of the signals for hydroxyl-bearing methines (3'-, 4'-, and 6'-H's), observed in the proton nuclear magnetic resonance (^{1}H-NMR) spectra upon acetylation, additionally suggested the sugar part to be β-*N*-acetylglucosaminide (β-glcNAc). Acid methanolysis of **1** yielded steroid **7** (in Scheme 1) and methyl D-glucosaminide. The latter was acetylated at the nitrogen and then con-

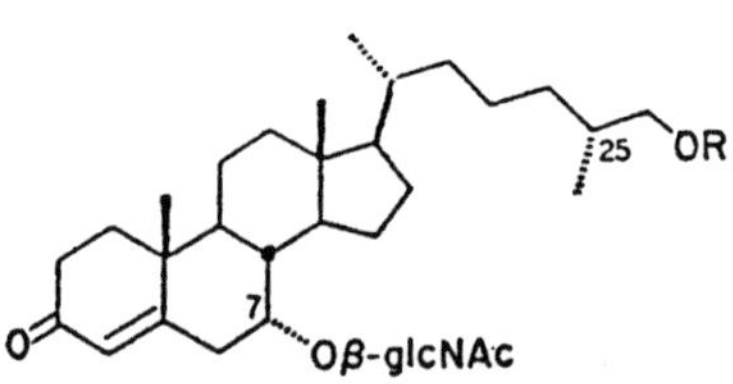

pavoninin-1 1 : R = Ac
" -2 2 : R = H

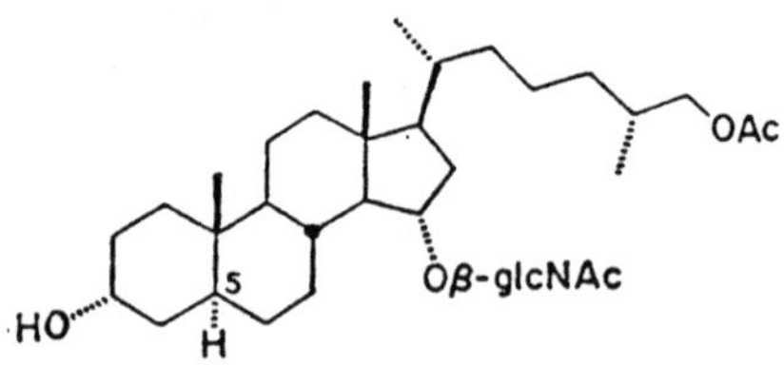

pavoninin-4 4

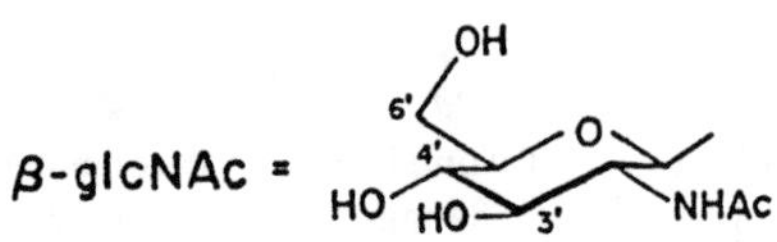

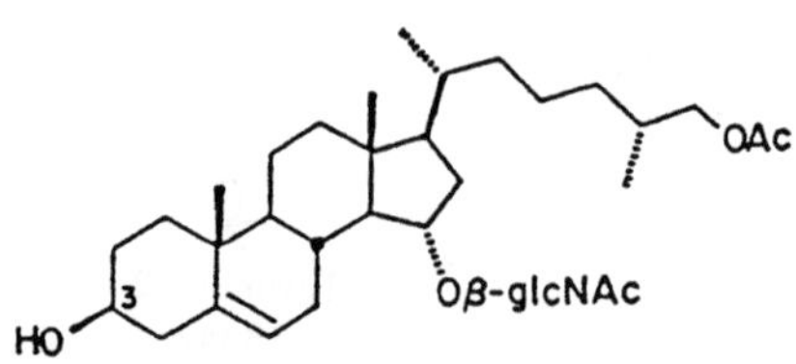

pavoninin-5 5

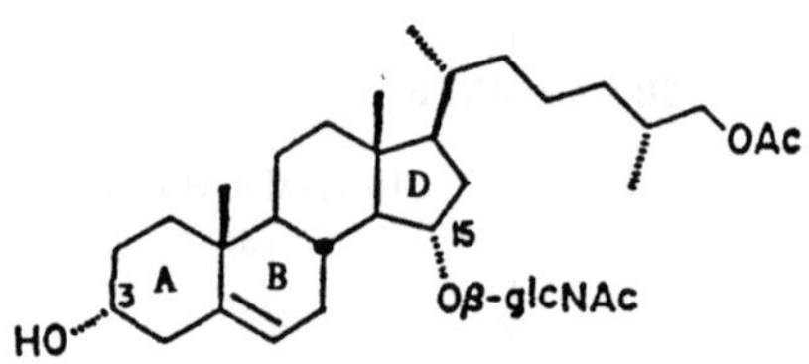

pavoninin-3 3

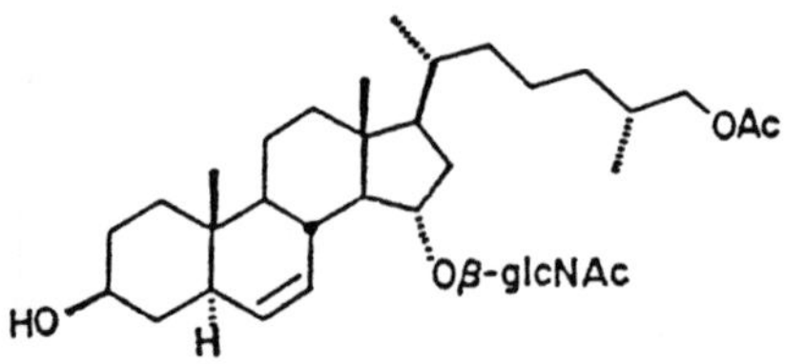

pavoninin-6 6

verted to the tri-*p*-bromobenzoate, the circular dichroic (CD) spectrum of which confirmed the sugar as glcNAc with absolute stereochemistry of D (Liu and Nakanishi 1981). The structure of steroid **7** was inferred by intensive double resonance ^{1}H-NMR measurements, and was established by chemical correlation with authentic cholesterol as shown in Scheme 1. Stereochemistry at the carbon where the sugar is attached (C-7) was determined by the ^{1}H-NMR spectrum of **1** that the oxygen is oriented axially, whereas that at the end of the side chain (C-25) was indicated by relative lanthanide-induced shifts of the methoxyl signals in ^{1}H-NMR between diastereomeric (+)- and (–)-2-methoxy-2-trifluoromethyl-2-phenylacetates (MTPA esters) of **7** (Yasuhara and Yamaguchi 1977). The latter stereochemistry was eventually established by synthesis of **8** (in Scheme 1) from authentic ergosterol and (2*S*)-3-chloro-2-methyl-1-propanol (Sakaitani and Tachibana, to be published).

The structure of pavoninin-2 (**2**) was evidently indicated by its spectroscopic data in comparison with those of **1**, and confirmed by peracetylation as yielding the peracetate identical to that from **1**. Suspicion remains that **2** is an artifact generated during lyophilization, because it was not found in the secretion squeezed from the dead sole and lyophilized independently, and also because slow solvolysis of the acetate moiety takes place when other pavoninins are left in solution of protic solvents. Be that as it may, **2** shows no weaker ichthyotoxicity than other pavoninins.

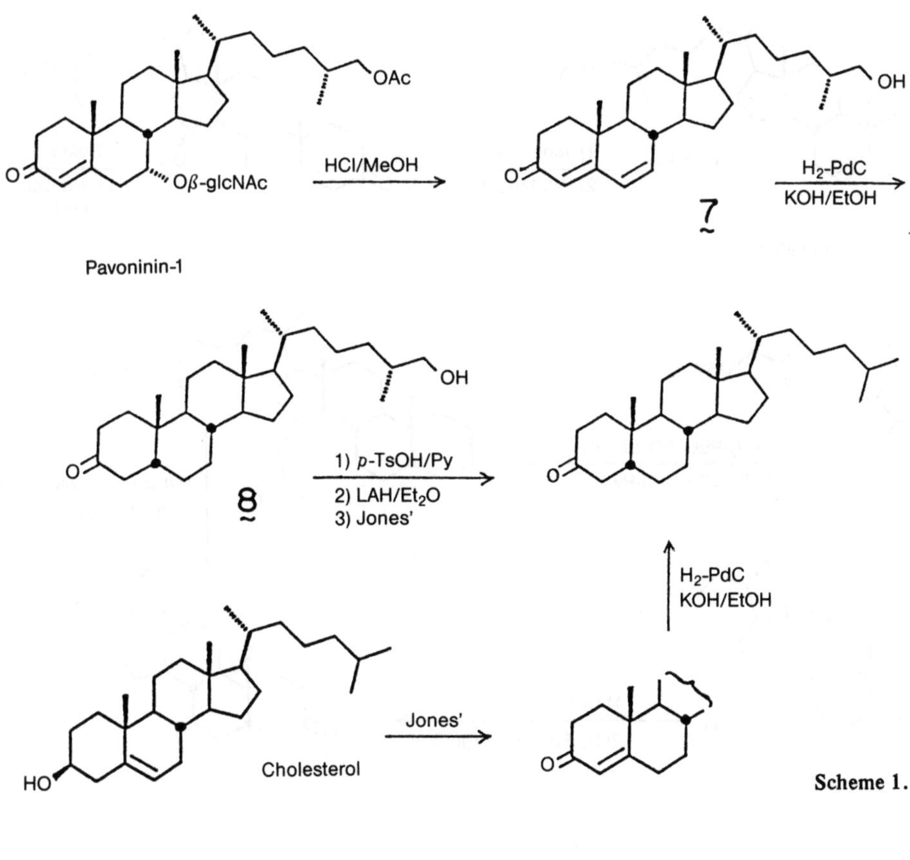

Scheme 1.

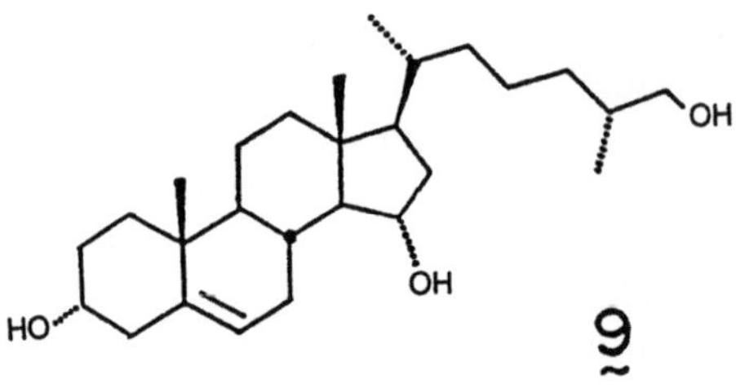

9

The spectroscopic data of pavoninin-3 (3) showed a somewhat different feature from 1 in that it has a hydroxyl on the steroidal frame instead of the ketone in 1, and the point where the sugar is attached is not in proximity to it. By a paramagnetic shift of the ^{1}H-NMR signal for the oxygenated methine in the A ring (C-3) upon acetylation, showing that a hydroxyl is there, the sugar was determined to be on the D ring (C-15). The acetylation shifts of the other three methines in the sugar also determined it to be β-glcNAc as in the case of 1. Intensive double resonance ^{1}H-NMR measurements on the steroid in the methanolysate of 3 suggested its structure as 9. Stereochemistry at C-3, i.e., the axial orientation of the hydroxyl, is based on its ^{1}H-NMR spectrum, whereas that at C-15 was determined by a CD spectrum of *p*-bromobenzoate of 10 (in Scheme 2) where the exciton chirality method is applicable to

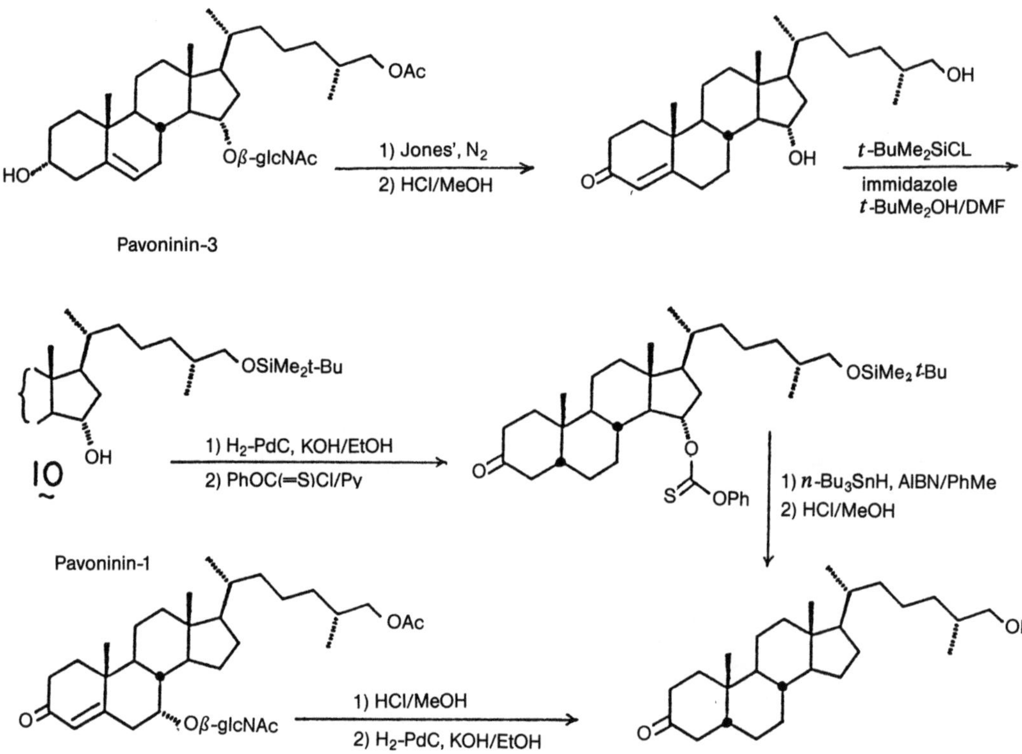

Scheme 2.

the two chromophores constituting a chiral coupled oscillator (Harada and Nakanishi 1983). **3** was chemically correlated to **1**, as shown in Scheme 2, thus establishing its entire stereochemistry.

Variation at the A/B ring moiety in the aglycons of pavoninins-4 (**4**), -5 (**5**) and -6 (**6**) from that of **3** was indicated by difference in their respective spectroscopic data. Namely, **4** was indicated to be a dihydro analog of **3** with 5α-hydrogen, and **5** to be the epimer of **3** at C-3. The structure of **6** was in turn indicated by spectroscopic comparison with **5**. They were chemically correlated mutually and to **3** as shown in Scheme 3, thus concluding chemical correlation among all the six pavoninins and cholesterol (The detailed description on the structure elucidation is to be published; Tachibana et al., Tetrahedron: in press).

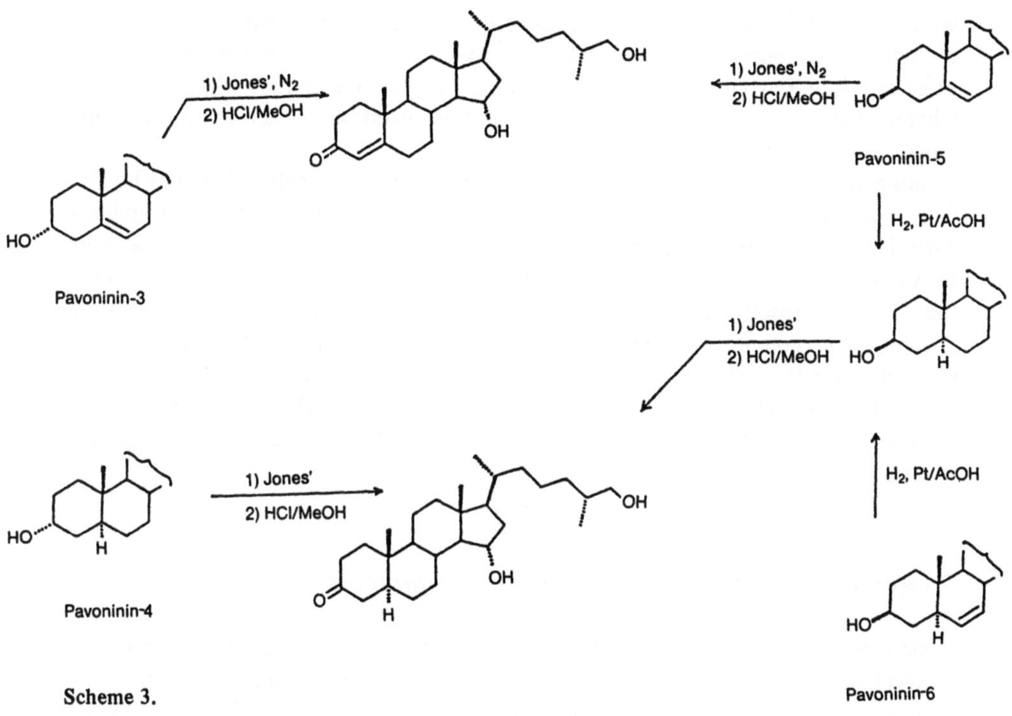

Scheme 3.

4 Shark Repellency Assays

In order to examine whether pavoninins, which had been isolated by tracing ichthyotoxicity, function as repellents against predatory fishes, qualitative assays were carried out with them as well as with the ichthyotoxic acetone precipitate on the dog shark *Mustelus grieseus* at the Okinawa Expo Aquarium. Imitating probable conditions of discharge by the sole, a stainless steel pipe (0.3 mm diam.) was connected to a 10-ml syringe at one end, and bent like a fishing hook at the other where bait fish was hooked, and put in the 500-l basin containing two sharks, each 40 cm long. Sample solutions were ejected when the hungry baited shark approached close to the hook, and the response of the shark was observed.

Control experiments were carried out with sea water and with 70% aqueous ethanol, when the shark simply bit off the bait as without any ejection. Escape behavior was clearly observed with pavoninins at the concentration of 0.9 mg ml^{-1} in the aqueous ethanol, but it was somewhat unclear at the higher concentration of 3.5 mg ml^{-1}. The same behavior was observed with standard saponin of 10 mg ml^{-1} sea water, but the reaction also lacked reproducibility as well as concentration dependency. These results can be explained that the repellency depends more on whether the ejection hits the shark's nostrils in this assay system, and close observations seemed to indicate that being the case. No reaction was shown by the shark with the acetone precipitate at the concentration of 100 mg ml^{-1} sea water, but sharks tended to spit out the bait when it was smeared with the precipitate beforehand. The bait smeared with pavoninins, on the other hand, tended to make sharks hesitate to attack it, though it was consumed immediately once they bit it. These observations together indicated that pavoninins are repellents which act on the shark's olfactory sense, whereas the acetone precipitate contains antifeedant(s) which act on its gustatory sense.

More recently, pavoninins were subjected to tonic immobility (TI) tests by Dr. Gruber at Miami (Gruber 1981; Gruber and Zlotkin 1982). According to preliminary TI tests on the lemon shark *Negaprion brevirostris,* the pavoninins have been shown to be relatively strong repellents by acting on buccal receptors as well as on the olfactory rosette at the concentration of 5 mg ml^{-1} in the aqueous ethanol (Gruber, pers. commun.). Considering the amount of pavoninins which the sole can eject at a single discharge *vide supra,* and the discharge being made at the last moment of the attack (Clark 1974), i.e., they reaching the shark's sense before much dilution, pavoninins can reasonably be presumed to function as repellents against sharks and other predatory fishes.

5 Lipophilic Toxins from Moses Sole *P. marmoratus*

Upon recognition of the repellent activity in pavoninins, a lipophilic fraction was correspondingly prepared from *P. marmoratus* secretion, and stronger ichthyotoxicity than pavoninins was recorded for it to the goldspot killifish *Floridichthys carpio* (carried out at Miami; Gruber pers. commun.). A portion of the fraction has thenceforth been subjected to preliminary chemical analyses. It comprises at least three major compounds which are a little less polar than pavoninins based on silica-gel tlc, and no compound with a UV-visible chromophore like pavoninin-1 is contained. Based on spectroscopic data presently available, a major compound appears to be also a steroid monoglycoside with a very similar aglycon to pavoninin-3 and a sugar, probably a monoacetylated pyranohexose other than glcNAc. The other components are also considered to be similar. Further investigation is in progress.

6 An Evolutionary Aspect of Pavoninins

Pavoninins are the first steroid glycosides characterized as defense substances of fish, although a number of ichthyotoxic saponins have been isolated from sea cucumbers and starfish (Hashimoto 1979). Existence of nonpeptidic defense substances, including pahutoxins, implies that ichthyocrinotoxic fish are not mere primitives of the venomous fish whose toxins are proteinaceous. In order to accomplish their objective as defense substances, ichthyocrinotoxins must work on the external senses of predators, whereas injected venoms must work in their internal physiological systems. Since the receptors of these toxins in their predators ought to differ, it is not rational to connect them by a continuous change of evolution. Ichthyocrinotoxic fish and venomous fish should therefore be considered as two different forms of fish furnished with defense substances, which have evolved independently through the history of natural selection.

Acknowledgments. I am grateful to Dr. T. Yoshino at Department of Marine Sciences, The Ryukyu University, for information and assistance on the collection of *P. pavoninus*, to Mr. M. Sakaitani, Department of Chemistry, Osaka University, for technical assistance in chemical conversions of pavoninins, to the instrumental analysis team of this Institute for spectroscopic measurements, to Mr. S. Uchida, Director, Okinawa Expo Aquarium, for assistance in the shark assay, and to Dr. S. H. Gruber, Rosenstiel School of Marine and Atmospheric Science, University of Miami, for TI tests. The favor of Dr. Gruber is also acknowledged together with that of Dr. E. Zlotkin, Department of Zoology, The Hebrew University of Jerusalem, for the supply of the lipophilic fraction of the *P. marmoratus* secretion. I am also indebted to Dr. K. Nakanishi, Director, this Institute, for direction and discussion throughout the present research.

References

Boylan DE, Scheuer PJ (1967) Pahutoxin: a fish poison. Science (Wash DC) 155: 52-56

Cameron AM, Endean R (1973) Epidermal secretions and the evolution of venom glands in fishes. Toxicon 11: 401-410

Clark E (1974) The Red Sea's shark proof fish. Natl Geogr Mag 146: 719-727

Clark E, George A (1979) Toxic soles, *Pardachirus marmoratus* from the Red Sea and *P. pavoninus* from Japan, with notes on other species. Environ Biol Fishes 4: 103-123

Goldberg AS, Wasylyk J, Renna S, Reisman H, Nair MSR (1982) Isolation and structural elucidation of an ichthyocrinotoxin from the smooth trunkfish (*Lactophrys triqueter* Linnaeus). Toxicon 20: 1069-1074

Gruber SH (1981) Shark repellents: perspectives for the future. Oceanus 24 (4): 72-76

Gruber SH, Zlotkin E (1982) Bioassay of surfactants as shark repellents. Nav Res Rev 34: 18-27

Halstead BW (1978) Poisonous and venomous marine animals of the world, revised edn. Darwin, Princeton, New Jersey

Harada N, Nakanishi K (1983) Circular dichroic spectroscopy – exciton coupling in organic stereochemistry. University Science Books, Mill Valley, California

Hashimoto Y, Oshima Y (1972) Separation of grammistins A, B and C from a soapfish *Pogonoperca punctata.* Toxicon 10: 279-284

Hashimoto Y, Shiomi K, Aida K (1974) Occurrence of a skin toxin in coral-gobies *Gobiodon* spp. Toxicon 12: 523-528

Hashimoto Y (1979) Marine toxins and other bioactive marine metabolites. Japan Scientific Societies, Tokyo

Hori K, Fusetani N, Hashimoto K, Aida K, Randall E (1979) Occurrence of a grammistin-like mucous toxin in the clingfish *Diademichthys lineatus.* Toxicon 17: 418-424

Liu H, Nakanishi K (1981) Additivity relation found in the amplitudes of exciton-split circular dichroism curves of pyranose benzoates. J Am Chem Soc 103: 5591-5503
Primor N, Zlotkin E (1975) On the ichthyotoxic and hemolytic action of the skin secretion of the flatfish *Pardachirus marmoratus* (Soleidae). Toxicon 13: 227-231
Primor N, Parness J, Zlotkin E (1978) Pardaxin: the toxic factor from the skin secretion of the flatfish *Pardachirus marmoratus* (Soleidae). In: Rosenberg P (ed) Toxins, animal, plant and microbial. Pergamon, Oxford, p 539
Yasuhara K, Yamaguchi S (1977) Use of shift reagent with MTPA derivatives in ^{1}H NMR spetroscopy. III. Determination of absolute configuration and enantiomeric purity of primary carbinols with chiral center at the C-2 position. Tetrahedron Lett: 4085-4088

Surfactant and Channel-Forming Activities of the Moses Sole Toxin

A. MORAN[1], Z. KORCHAK[2], N. MORAN[3], and N. PRIMOR[2]

1 Introduction

Among the group of flatfishes (Order Pleuronectiformes) only a few species within the *Pardachirus* genus are known to possess toxins in specialized glands located along their dorsal and anal fins (Clark and George 1979). Since it was demonstrated that the gland secretion repels sharks, a function of a self-defense mechanism for a flatfish toxin was suggested (Clark 1974, 1983). Studies have shown a diversity of effects produced by the toxic secretion or by its isolated toxin (pardaxin) in biological membranes from various origins, including viruses (Pal et al. 1981 a, b), red blood cells (Primor and Lazarovici 1981), nervous tissue (Spira et al. 1976; Parness and Zlotkin 1976; Primor 1984) and iontransporting epithelia (Primor 1983). Its histopathological effects seem to be associated with its surfactant-like property (Primor et al. 1983; Zlotkin and Barenholz 1983) and with an increase in membrane permeability.

In the present work, the planar lipid bilayers were used as a model system to study the nature of interaction between the flatfish toxin and biological membranes. The results of the study support our original hypothesis (Moran et al. 1978) that it acts on the lipid bilayer portion of the cell membrane to produce transmembranal channels which cause increase in membrane permeability.

The surfactant-like quality of the toxin enhances its interaction with the membrane. A preliminary version of this study has appeared in an abstract form (Moran et al. 1978).

Present address:
1 Armed Forces Radiobiology Research Institute, Department of Physiology, Bethesda, MD 20814, USA
2 Osborn Laboratories of Marine Sciences, New York Aquarium New York Zoological Society, Brooklyn, NY 11224, USA
3 Biophysics Lab, NINCDS, Bethesda, MD 20205, USA

Toxins, Drugs, and Pollutants in Marine Animals
(ed. by Bolis et al.)

2 Materials and Methods

2.1 Assays of Toxicity in Fish

Killifish *(Fundulus heteroclitus)* of both genders were collected from the estuaries near the Osborn Laboratories of Marine Sciences, New York Aquarium, Coney Island, Brooklyn, New York. They were adapted to various concentrations of artificial sea water (Utility Chemical Co., Paterson, N.J. USA) for a period of 3 weeks and fed shrimp and squid daily. These fish can tolerate a very wide range of salinities (Karnaky et al. 1976). The salinities of the media were determined by conductivity measurements (Conductivity meter, Presto-Tek Co., Los Angeles, CA, USA) and compared to Standard Sea Water (Standard Sea Water Service, Charlottenlund, Denmark) which has 19.38°/oo chlorinity and 35.03°/oo salinity, and expressed as percent of Standard Sea Water (SSW). The isotonic medium of adaptation consists of about 17% of SSW (Potts and Oates 1983).

Benzalkonium chloride, sodium dodecyl sulfate and pardaxin were administered to the fish in two modes: (1) In the adaptation media; 4 groups of 10 fish (each weighing 2.0 ± 0.3 g) were introduced into glass containers containing 500 ml of one of the adaptation media and different concentrations of the toxic substance. The pH of the media was adjusted to 7.2–7.3 before the fish were introduced. (2) By injection; the tested substances were dissolved in a fish Ringer's solution (Degnan et al. 1977). Then 2.5μl, 5μl, 10μl, or 20μl of the solution was injected into the peritoneal cavity of each of the 10 fish (each weighing 2.0± 0.3 g), which had been adapted to 95% salinity of SSW. Lethality, based on complete immobilization, was determined after 1 h. The lethal dose (LD_{50}) was calculated by the method of Reed and Muench (1938).

2.2 Contact Angle

The contact angle (θ) is a quantitative manifestation of the surface tension activity of the substance (Absolom et al. 1981). To determine the surfactant charateristics of the principal toxin (pardaxin) of the *P. marmoratus* secretion, it was mixed with saline. Then 10 μl drops of the saline solution containing various concentrations of paradaxin were applied with a mircopipette to the solid surface of a hydrophobic polystyrene film (Central Research Lab., Dow Chemical, Midland Michigan, USA). The contact angle from drop profiles was taken as the angle between the tangent to the drop where it touches the surface and the surface itself. The contact angles were measured using a small telescope with a cross-hair, attached to a goniometer (Model 100, NRL Contact Angle Goniometer, Rame-Hart, Inc., Mountain Lakes, NJ, USA).

Measurements were taken at room temperature (20°–22°C), within 15–20 s after a fairly constant angle was established. Five measurements were taken, from different drops, for each pardaxin concentration, and the results are given as a mean ± S.D. Pardaxin was dissolved in buffer solution (saline) composed of (in g 1^{-1}): 8.0 NaCl, 0.2 KCl, 1.15 Na_2PO_4, 0.2 KH_2PO_4, and 1.0 Na_2EDTA.

2.3 Membranes and Electrical Measurements

Lipid bilayer membranes were made from solution of 2% soybean-lecithin in n-decan. The lipid solution was applied with a syringe to a small hole (0.017 cm^2) punched in a Teflon septum separating two halves of a lucite chamber. The aqueous solution was composed of various concentrations of KCl in 5 mM Tris, pH 7.4. The cell and the measuring system were similar to those described previously (Moran and Ilani 1974). Electrical contact to each side of the bilayer was made with a calomel electrode via salt bridge (saturated KCl in 3% agar.) Membrane formation was monitored by visual appearance and electrical capacitance criteria. Membrane resistance was measured under voltage clamp conditions. Following a change in potential, the membrane current assumed its steady-state value within a few seconds. Therefore, steady-state current-voltage relationships could be obtained over a wide range of voltages by clamping the membrane potential with a slow (0.01 cps) ramp and feeding directly both voltage and current to two inputs of an X–Y plotter (Yokogawa 3047, Japan). The unmodified lipid bilayers displayed electrical conductance from 5.6×10^{-9} S cm^{-2} to 2.8×10^{-10} S cm^{-2} and electrical capacitances of about 0.5 μF cm^{-2}. Only membranes with conductances of less than 10^{-9} S cm^{-2} were used.

2.4 Chemicals

Pardachirus marmoratus fish (Pisces; Soleidae) were collected from the Red Sea (Eilat, Israel). The toxic secretion was obtained by the method described by Clark and Chao (1973) and stored in lyophylized form. A stock solution of 10^{-3} g ml^{-1} was prepared in distilled water and kept at −20 C^{o}. This solution was stable for a period of 4 to 5 months with no measurable loss of its activity on lipid membranes.

Pardaxin, the isolated proteinaceous toxic component, was obtained according to Primor et al. (1978). The molecular weight of the reduced and alkylated pardaxin was estimated by SDS gel electrophoresis, where it produces a single band corresponding to M.W.=13,100 (Primor, in preparation).

The concentrations of benzalkonium chloride (BAC), sodium dodecyl sulfate (SDS) and pardaxin are given as molar (mol l^{-1}). For calculations of molar ratios the following molecular weights were assumed: benzalkonium chloride, 360; sodium dodecyl sulphate, 288.38; pardaxin, 13,000: Soybean lecithin was purchased from Sigma Chemical Co. (St. Louis, MO, USA); cleaned by washing with acetone n-decan (Fluka Chemical Corp., Hauppauge, N.Y., USA); Benzalkonium chloride (Ruger Chemical Co., Irvington, N.J., USA). Other chemicals were of analytical grade. Water was double-distilled.

3 Results

3.1 Assay of Lethality to Fish of BAC, SDS and Pardaxin

Previously, pardaxin (PX) was shown to be more toxic to *Aphanius dispar* fish when dissolved in the medium than when injected into the fish. In addition, its toxicity was greater in hypo- and hyper-osmotic media than in media isotonic to the fish's body fluid (Primor et al. 1980; Primor et al. 1983; Primor 1983).

The toxicity of BAC, SDS, and PX were tested in *F. hetereoclitus* by injection to fish and by administration to the media of adaptation. BAC and SDS were about equally toxic in 180% SSW, while BAC was more toxic at the other salinities with ratios of 2.1, 4.2, and 6.2 for fish in 95%, 17%, and 0.008% of SSW, respectively. PX was toxic at much lower concentrations, about 10 times more toxic (based on molar concentration) than BAC and SDS in media of 180% SSW, but in 0.008% SSW, PX was 350 times more toxic than SDS (Table 1).

Table 1. Assays of toxicity[a] of benzalkonium chloride (BAC), sodium dodecyl sulphate (SDS) and pardaxin (PX) in *F. heteroclitus* fish adapted to media of various concentrations of sea water

	Media of adaptation (% of the Standard Sea water)[b]							
	180		95		17		0.008	
	M	mg ml^{-1}	M	mg ml^{-1}	M	mg ml^{-1}	M	mg ml^{-1}
BAC	1.5×10^{-5}	0.0055	5.8×10^{-5}	0.021	4.2×10^{-5}	0.020	4.2×10^{-5}	0.015
SDS	1.5×10^{-5}	0.0045	9.0×10^{-5}	0.026	1.8×10^{-4}	0.052	2.7×10^{-4}	0.080
PX	3.8×10^{-7}	0.005	2.7×10^{-6}	0.036	5.0×10^{-6}	0.066	7.6×10^{-7}	0.010

[a] The data are expressed in units of molar and milligram of the toxic substance per milliliter of the adaptation media which elicited death in 50% of the tested fish (LD_{50}). All substances were given to the media and each LD_{50} was determined using 4 concentrations of the tested substance in 500 ml of the adaption media and 10 fish (weighing 2 ± 0.3 g each). Lethality based on a complete immobilization was determined following 1 h.

[b] Prior to the assays of toxicity the killifish were allowed to adapt to the various media for a period of 3 weeks.

BAC, SDS and PX were significantly less toxic when injected into the fish than when added to the media externally (Table 1). Lethal dose (LD_{50}) of 1.5, 6.0 and 2.8 mg were obtained, respectively, using fish weighing 2g. adapted to 95% SSW.

3.2 Effect of Pardaxin on the Contact Angle of Drops of Saline

There was a dose-related correlation between the concentration of PX and the contact angle (θ). The latter varied from 96° (the control value) to 42° at 1.1×10^{-4}M PX. At concentrations higher than 1.5×10^{-5}M, the drop lost its sharp ball like profile and began to flatten out. This reduced the accuray of the measurements of

Table 2. Effect of pardaxin (PX) on the contact angle (θ) of drops of saline

Control	PX concentration (M)	Contact angle (θ) [a] (degrees)	% of decrement
		96 ± 1	
	3.2×10^{-6}	95 ± 2	0
	5.1×10^{-6}	82 ± 3	14.5
	7.5×10^{-6}	76 ± 3	20.8
	1.1×10^{-5}	65 ± 5	32.3
	1.5×10^{-5}	58 ± 7	39.6
	1.8×10^{-5}	51 ± 7	46.9
	3.6×10^{-5}	45 ± 9	53.1
	1.1×10^{-4}	42 ± 11	56.2
	2.2×10^{-4}	42 ± 11	56.2

[a] Drops of 10 μl were placed on a hydrophobic polystyrene film. Measurements were taken from the drop profiles of an angle which is formed between the drop and the solid surface at the place where the drop touches that surface. The data are mean ± S.D. of five measurements.

the contact angle, which is reflected in the high standard error of the mean seen in Table 2. A similar phenomenon ist discussed in detail by Huh and Reed (1983). The concentration of PX which produced 50% decrease in contact angle was 9.5×10^{-6} M as estimated from the dose-response curve. The lowest concentration of PX which produced a detectable effect on the contact angle was 5.1×10^{-6} M (Table 2).

3.3 Effects of Pardachirus Toxic Secretion on Lipid Membranes

The addition of 1.1×10^{-7} g ml^{-1} of *Pardachirus* toxin to the aqueous compartment on one side of the membrane (cis) increased its conductance about 100-fold. Rinsing of the chamber from the toxin caused a gradual decrease of the membrane conductance from 10^{-8} S cm^{-2} to 10^{-9} S cm^{-2} within 20 min. However, in spite of repeated wash during 90 min the original resistance has not been completely restored (Fig. 1).

Consecutive additions of the toxin to the cis compartment under voltage clamp conditions result in rapid drops in membrane resistance, each time to a new lower value of steady-state resistance. Plotting the toxin concentration and membrane resistance on logarithmic scales reveals a linear relationship (Fig. 2). The slope of this line (between 6 and 8) indicates that an aggregate of 6 to 8 molecules of the toxin is needed to form an active form which is able to affect the membrane conductance (Blumenthal and Klausner 1982).

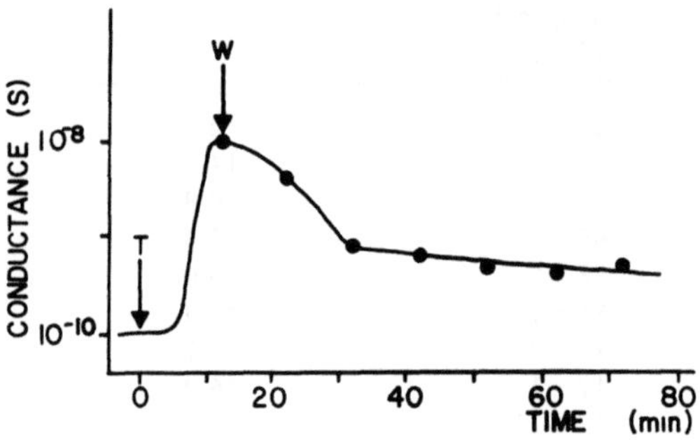

Fig. 1. The effect of *P. marmoratus* toxin on conductance of planar lipid bilayer as a function of time. A membrane formed from 2% soybean-lecithin in n-decane separated symmetrical salt solutions (100 mM KCl, 5 mM Tris, pH 7.4). The toxin (T) at 1.1×10^{-7} g ml^{-1} was added to one side of the lipid bilayer. Following an increase in conductance to its steady-state value the procedure of washing out (W) of the toxin begins. Note that most of the effect is reversible although the lipid membrane remains at lower resistance in spite of prolonged was-out

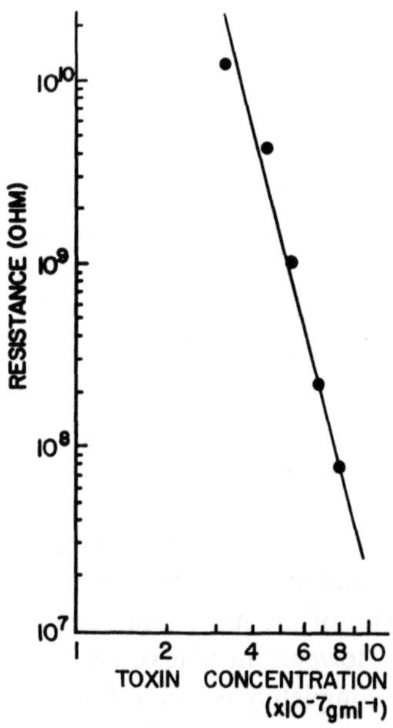

Fig. 2. Double logarithmic plot of membrane resistance vs. concentration of *P. marmoratus* toxin. A membrane formed from 2% soybean-lecithin in n-decane separated symmetrical salt solutions (100 mM KCl, 5 mM Tris, pH 7.4). Successive additions of toxin were made to the solution on one side (cis) of the membrane. The points represent values of steady-state resistance measured at membrane potential of 50 mV (positive at the cis side)

When the potential is negative with respect to the side without the toxin (trans), the current-voltage relationship remains linear, as in the nontreated membrane. However, when the potential is reversed, i.e., is made positive with respect to the trans side of the membrane, a nonlinear behavior is observed, Above a certain voltage the current increases steeply (Fig. 3, line A). We defined the threshold as the voltage at which the chord conductance (indicated in Fig. 3 by the broken line) equals twice the ohmic conductance at the opposite potential.

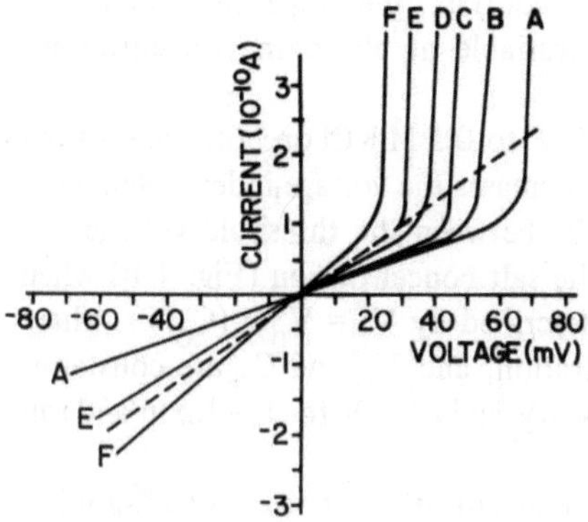

Fig. 3. Steady-state current-voltage (I-V) relationship of a membrane treated with different concentrations of *P. marmoratus* toxin. Membrane formed from 2% soybean-lecithin in n-decane separated symmetrical salt solutions (100 mM KCl, 5 mM Tris, pH 7.4). Toxin was added to one side of the membrane (cis). *Solid lines: A* 2.8 × 10^{-7} g ml^{-1}; *B* 4.2 × 10^{-7} g ml^{-1}; *C* 5.7 × 10^{-7} g ml^{-1}; *D* 7.2 × 10^{-7} g ml^{-1}; *E* 8.5 × 10^{-7} g ml^{-1}; *F* 10^{-6} g ml^{-1}. *Broken line:* chord conductance equal to twice the ohmic conductance obtained with 2.8 × 10^{-7} g ml^{-1} of toxin. At this concentration of toxin the threshold voltage is the membrane potential at which the broken line intersects with the solid line A. Polarity of voltage corresponds to the cis side of the membrane. Note that the toxin changes the ohmic conductance and the threshold voltage

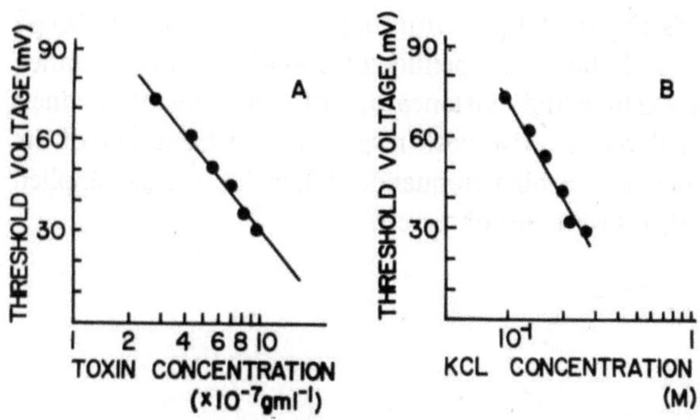

Fig. 4 A, B. Effect of concentration of *P. marmoratus* toxin **(A)** and of KCl **(B)** on threshold voltage. Membrane formed from 2% soybean-lecithin in n-decane separated symmetical salt solutions. Toxin was added to one of the membrane. **A** Points taken from experiments similar to those given in Fig. 3. **B** 2.8 × 10^{-7} g ml^{-1} of toxin was added to solutions of different KCl concentrations (in 5 mM of Tris, pH 7.4)

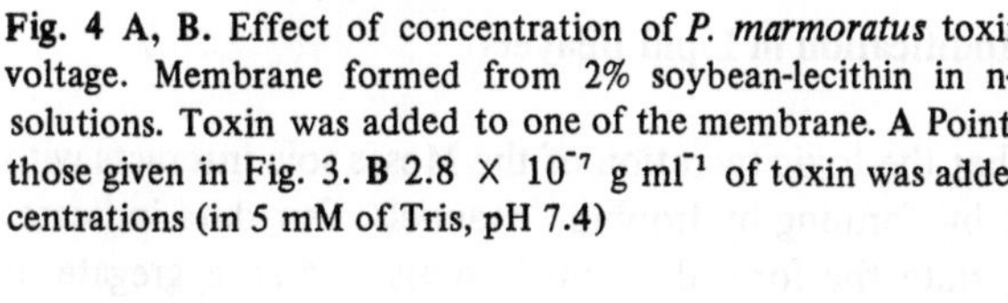

Fig. 5. Amplitude histogram of conductance fluctuations in lipid membrane treated with *P. marmoratus* toxin. Membrane formed from 2% soybean-lecithin in n-decane separated symmetrical salt solutions (2 M KCl, 5 mM Tris, pH 7.4). Toxin (5.6 × 10^{-9} g ml^{-1}) was added to one side (cis) of the membrane. The voltage was clamped at a potential of 60 mV (cis side positive). *Inset* the fluctuations of membrane current

An increase of the toxin concentration has two distinct effects: (1) it increases the voltage-independent (ohmic) conductance of the treated membrane and (2) it decreases the threshold voltage (Fig. 4 A). In contrast, the slope conductance at potentials beyond threshold is practically invariable at all toxin concentrations (Fig. 3).

An increase of salt concentration from 0.05 M to 0.5 M KCl on both sides of the membrane decreases the threshold voltage and increases the voltage-independent conductance (Fig. 4 B). A linear relationship results between the threshold voltage and either the toxin concentration (Fig. 4 A) or the salt concentration (Fig. 4 B) when plotted on semi-logarithmic scales. It can be described by $V_t = V_0 \ln (C_0/C)$, where V_t is the threshold voltage, C is the concentration, and V_0 and C_0 are constants. V_0 from the slopes of the lines in Fig. 4, is 34 mV and 45 mV for the toxin and salt concentration dependence, respectively.

When the membrane is treated with a low concentration of toxin (10^{-9} g ml^{-1}) fluctuations of current appear under voltage clamp conditions, corresponding to the opening and closing of individual single channels in the lipid membrane. At 2M KCl the single channel conductance is about 250 pS, with multiples occurring at 500 pS and 1000 pS, possibly due to simultaneous opening of more than one channel, although at present we cannot exclude the existence of different sizes of channels (Fig. 5). When the voltage applied was positive with respect to the trans side of the membrane, the current jumps occured at high frequency. When the voltage applied was negative, the current fluctuations were not observed.

4 Discussion

4.1 Mechanism of Membrane Modification in Lipid Bilayers

The present study has shown that the toxic secretion of the Moses sole interacts with an artificial bilayer membrane by forming hydrophilic channels. Our data indicates that at the highly conducting state the formed channel consists of an aggregate of 6–8 monomers. By comparison, in the low conducting state, i.e., the "ohmic" region of I–V relationship (Fig. 3), the conducting unit seems to consist only of single monomers (using the criterion of the slope of log ohmic conductance vs. log concentration of toxin (not shown), Blumental and Klausner 1982).

This and the observation that the frequency of opening of single channels increases sharply above the threshold potential (Fig. 5) indicate that the threshold is the voltage at which enhanced aggregation of monomers occurs. Since the membrane-active factor (pardaxin) is negatively charged in neutral pH (Primor and Zlotkin 1976), and since applied potential needs to be positive on the toxin side of the membrane for the cooperative interaction to occur, it is conceivable that the effect of voltage is to align the monomers near and/or in the membrane so that cooperative interaction or aggregation is facilitated. Raising the salt concentration probably enhances the aggregation of the PX molecules by diminishing electrostatic repulsion between the similarly charged particles (Kloppenol 1980). A similar phenomenon has been described for other channel-forming agents in lipid bilayers such as alamethicin and related analogues (Boheim et al. 1978). The accumulative effect of insertion of channels into the bilayer causes ultimately the breakage of the membrane. We observed this effect over the range of concentration from 10^{-9} g ml^{-1} (where single channels were recorded) to 10^{-7} g ml^{-1} (where breakage occured). Similar phenomenon most probably occurred during lysis of liposomes (Zlotkin and Barenholz 1983). This rather narrow range of concentration for PX's action is consistent with high aggregation number (6-8) for the formation of the channel (Fig. 2).

4.2 Surfactant Character of Pardaxin

We would like to emphasize that the studies on the lipid bilayers presented here and the observation that PX reduces the size of saline drops were the basis for the theory that it may act directly on lipid bilayer via its surfactant properties and through formation of channels, without the need to interact with a specific receptor on the surface of the membrane (Moran et al. 1978).

The steep decrease of contact angles (θ) measured from the profiles of drops containing PX is a direct proof for its surfactant quality. Pardaxin has the highest surface-active potency of any protein measured so far. For example, human serum albumin at 3.5 mg ml^{-1} (10^{-5}M) reduced the contact angle from 86° to 72° (Absolom et al. 1981) which corresponds to 16.3% decrement. Pardaxin, however, exerted a similar effect at a concentration of 6×10^{-6}M, i.e, about ten times more dilute, and its maximum percentage of decrement was 56.2% (Table 2). It is not unique for biologically active substances to be armed with surfactant quality. For example, proteins with a surfactant characteristis secreted by specific alveolar cells in the lung, provide low surface tension at the air-liquid interface and this facilitates gas exchange (Dobbs et al. 1982; Magoon et al. 1983). Although spider- and tetanus- and diphtheria-toxins (Finkelstein et al. 1976; Deleers et al. 1983; Boquet and Duflot 1982) form channels, these toxins are not known to have repellent effect and their biological action is not due to their surfactant property to our knowledge. Yet, an example combining the tow qualities: surfactant and channel-forming activities, is a toxin isolated from a sea anemone (Ash et al. 1981; Michaels 1979; Shin et al. 1979; Varanda and Finkelstein 1980).

Some surfactants such as Trition X-100 and SDS are known to act on the lipid bilayer to form channels (Schlieper and DeRobertis 1977; Abramson and Shamoo 1979). Thus, due to its amphipatic character (Zlotkin and Barenholz 1983), PX may insert its hydrophobic part into the lipid bilayer to form a transmembrane channel by the aggregation of several molecules as observed in the artificial lipid bilayer. Or, on the other hand, the toxin can behave as detergent, i.e., an agent simply disordering the membrane matrix. In addition, in biological membranes interactions between the toxin and a specific receptor could take place. BAC, SDS, and PX display one similarity: they were less toxic when injected into fish than when dissolved in the adaptation media. It is difficult to estimate the concentration of the injected surfactants in the fish's body. If one assumes that these substances disperse uniformly in the body of a 2-g fish, the LD_{50} values, expressed in units of concentration within the fish, will correspond to 0.75, 3.0, and 2.4 mg ml^{-1} for BAC, SDS, and PX, respectively. This approximation reveals that BAC, SDS, and PX were 35, 115, and 67 times less toxic, respectively, when injected into the fish than when applied externally (compared with Table 1, column of 95% of SSW). These findings suggest that these substances act upon the outer surface of the fish. Indeed, previous studies have suggested that in fish, the gills are the most likely target for PX's action (Primor 1983). We do not know why this target organ seems to be preferably accessible for the luminal side. This may reflect a specific interaction with membrane receptors, prior to the formation of channels. More likely, however, it may be due to the specific conditions existing on the luminal side, such as acidity, salinity, potential difference, etc. These possibilities should be further tested.

4.3 Ichthyotoxicity and Noxious Effects vs. Surfactant Activity

Interestingly, PX's LD_{50} in fish maintained at 0.008% of SSW was at 7.6×10^{-7}M (Table 1). At this low concentration PX has little effect as a surfactant, as measured by the contact angle method. It is quite possible, therefore, that its action on living cells is enhanced even at this low concentration by the transmembrane potential difference (positive on the toxin containing side of the membrane) and its major mode of action at least at this concentration is via formation of channels. At 5×10^{-6}M PX displayed a distinct surfactant effect as well as toxicity to fish in an isotonic medium of adaptation (17% of SSW, Table 1). In isotonic medium PX produces a net sodium entry in the gill-like operculum epithelium (Primor 1983). Since increasing salt concentration enhanced the channel information mechanism in the lipid bilayers (Fig. 4 B); one may speculate that this effect adds to PX toxicity in fish living in a medium of higher salinity and channel forming constitutes the major determinant of toxicity even at higher concentration of PX.

Pardaxin and SDS have been shown to repel sharks (Gruber and Zlotkin 1982; Gruber 1982; Primor et al. 1984). However, several other tested potent surfactants had no repellent effects even at a concentration 100 times higher than SDS (Gruber and Zlotkin 1982). This finding suggests that surface activity alone is not a sufficient prerequisite to induce the repellent effect in sharks. Pardaxin and SDS share a common action on membranes: they both produce conducting channels in lipid membranes (this Vol.; Abramson and Shamoo 1979) and elicits a net sodium flux by acting predominantly on the luminal side in epithelial tissues. (Primor 1983; Sund and Olsen 1981). In addition, a peptide from dart frog known to have noxious effects also forms channels and it, too, acts on the luminal site only in ion-transporting epithelia (Greenwell and Low 1981); Montecucchi and Henschen 1981). Although this effect is limited to the luminal side of the epithelia, the diversity of the agents renders the specific receptor hypothesis unlikely. Rather, this common action exerted by entirely different chemicals suggests that the repellent effect is produced by those chemicals which act on the luminal side of epithelia, from transmembrane channels and produce a net sodium flux. This hypothesis should be further tested.

Acknowledgments. Financial support from: The Department of Defense, Office of Naval Research, Research Grant No. N00014-82-C-0435 is acknowledged. Many thanks to Rame-Hart, Inc., Mountain Lakes, New Jersey for allowing measurements of contact angles; and to Dow Chemical Corp., Midland, Michigan, for generous gift of polystyrene films

References

Abramson JJ, Shamoo AE (1979) Anionic detergents as divalent cation ionophores across black lipid membranes. J Membr Biol 50: 241–255

Absolom DR, Van Oss CJ, Zingg W, Neumann AW (1981) Determination of surface tensions of proteins. Biochim biophys Acta 670: 74–88

Ash P, Hider RC, Menez A, Wunderer G (1981) Surface activity of polypeptide isolated from *Anemonia sulcata.* Biochim biophys Acta 669: 231–235

Blumental R, Klausner RD (1982) The interaction of proteins with black lipid membranes. In: Posta G and Nicolson GL (eds) Membrane Reconstitution. Elsevier Biomedical, New York, p 43

Boheim G, Inmscher G, Jung G (1978) Trichotoxin A–40, A new membrane – exciting peptide. Biochim biophys Acta 507: 485–506

Boquet P, Duflot E (1982) Tetamus toxin fragment forms channels in lipid vesicles at low pH. Proc Natl Acad Sci (USA) 72: 7614–7618

Cancalon P (1983) Influence of detergent on the catfish olfactory mucosa. Tissue & Cell 15: 245–258

Clark E (1974) The Red Sea's shark proof fish. Nat Geogr Mag 146: 718–728

Clark E (1983) Repellent effect of the Red Sea moses sole. In: Zahuranec BJ (ed) Shark repellents from the sea: new perspectives. American Association for the Advancement of Science, Washington DC, p 1-35

Clark E, Chao S (1973) A toxic secretion from the Red Sea flatfish *Pardachirus marmoratus* Lacepede). Sea Fish Res Stn Haifa Bull 60: 53–56

Clark E, George A (1979) Toxic soles, *Pardachirus marmoratus* from the Red Sea and *P. pavoninus* from Japan with notes on other species. Environ Biol Fishes 4: 103–123

Degnan KJ, Karnaky KJ, Zadunaisky JA (1977) Active chloride transport in the *in vitro* opercular skin of a teleost *(Fundulus heteroclitus)*, a gill-like epithelium rich in chloride cells. J Physiol (Lond) 271: 155-191

Deleers M, Beugnier N, Falmagne P, Cabiaux V, Ruysschaert JM (1983) Localization in diphteria toxin fragment B of a region that induces pore formation in planar lipid bilayers at low pH. FEBS Lett 160: 82–86

Dobbs G, Mason RJ, Williams MC, Benson BJ, Sulishi K (1982) Secretion of surfactant by primary cultures of alveolar type II cells isolated from rats. Biochim Biophys Acta 713: 118–127

Finkelstein A, Rubin LL, Tzeng MC (1976) Black widow spider venom: effect of purified toxin on lipids bilayer membranes. Science wash DC: 1009–1011

Greenwell JR, Low MS (1981) Action of caerulein, gastrin 17, pentagastrin, and secretion on the active transport of sodium by the frog skin. J Membr Biol 61: 71–96

Gruber SM, Zlotkin E (1982) Bioassy of surfactants as shark repellents. Nav Res Rev 34: 18–27

Gruber SM (1982) Shark repellents: perspectives for the futurs. Oceanus 24: 72–76

Huh C, Reed RL (1983) A method for estimating interfacial tensions and contact angles from sessile and pendant drop shapes. J Colloid Interface Sci 91: 472–484

Karnaky KJ, Kinter LB, Kinter WB, Stirling EC (1976) Teleost chloride cell. Autoradiographic localization of gill Na, K-ATPase in killifish *Fundulus heteroclitus* adapted to low and high salinity environments. J Cell Biol 70: 157–177

Koppenol WH (1980) Effect of molecular dipole on the ionic strength dependence of a bimolecular rate constant. Identification of the site of reaction. Biophys J 29: 493–508

Magoon MW, Wright JR, Baritussio A et al. (1983) Subfractionation of lung surfactants implication for metabolism and surface activity. Biochim Biophys Acta 750: 18–31

Michaels DW (1979) Membrane damage by a toxin from the sea anemone *Stoichactis helianthus*. Formation of transmembrane channels in lipid bilayers. Biochim Biophys Acta 555: 67–78

Montecucchi PS and Henschen A (1981) Amino and composition and sequence analysis of sauvagine, a new active peptide from the skin of *Phyllomedusa sauvagei*. In J Peptide Protein Res 18: 113–120

Moran A and Ilani A (1974) The effect of prymmesium toxin on the electrical conductivity of thin lipid membrane. J Membr Biol 16: 237–256

Moran A, Primor N, Zlotkin E (1978) The effect of *Pardachirus* venom on the conductance of lipid bilayers. Isr J Med Sci 14: 499–500 (abstract)

Pal R, Barenholz Y, Wagner RR (1981a) Pardaxin a hydrophobic toxin of the Red Sea Flatfish, disassembles the intact membrane of vesicular stomatitis virus. J Biol Chem 256: 10209–10212

Pal R, Barenholz Y, Wagner RR (1981b) Transcription of vesicular stomatitis virus activated by pardaxin, a fish toxin that permeabilized the virion membrane. J Virol 39: 671–645

Potts WTW and Oates K (1983) The ionic concentrations in the mitochondria rich or chloride cell of *Fundulus heteroclitus*. J exp Zool 227: 349–360

Parness J, Zlotkin E (1976) Action of the toxic secretion of the flatfish *Pardachirus marmoratus* on the guinea pig ileum. Toxicon 14: 85–91

Primor N, Zlotkin E (1976) The toxic and antitoxic factors derived from the skin secretion of the flatfish *Pardachirus marmoratus* (Soleidae). In: Ohsaka A, Hayashi K and Sawai Y (eds) Animal, plant, and microbial toxins. Plenum, New York, p 287

Primor N, Parness J, Zlotkin E (1978) Pardaxin: the toxic factor from the skin secretion of the flatfish *Pardachirus marmoratus* (Soleidae). In: Rosenbert P (ed) Toxin, animal, plant and microbial. Pergamon, Oxford, p 539

Primor N, Lazarovici P (1981) *Pardachirus marmoratus* (Red Sea flatfish) secretion and its isolated toxic fraction pardaxin: the relationship between hemolysis and ATPase inhibition. Toxicon 19: 573–578

Primor N, Sabmay I, Lavie V, Zlotkin E (1980) Toxicity to fish, effect on gill ATPase and gill ultrastructural changes induced by *Pardachirus* secretion and its derived toxin pardaxin. J Exp Zool 211: 33–43

Primor N (1983) Pardaxin produces sodium influx in the teleost gill-like operculum epithelia. J Exp Biol 105: 83–94

Primor N, Zadunaisky JA and Tu AT (1983) Pardaxin, its action, toxicity and structure. In: Zahuranec BJ (ed) Shark repellents from the sea: New perspectives. American Association for the Advancement of Sciene, Washington DC, p 173

Primor N, Zadunaisky JA, Murdaugh HV, Boyer JL, Forrest JN (to be published 1984) Pardaxin increases solute permeability of gills and rectal gland in the dogfish shark *(Squalus acanthias)*. Comp. Physiol. Biochem. Part C

Primor N (1984) Pardaxin produces postjunctional muscle contraction in guineapig intestinal smooth muscle. Br J Pharmacol 82: 43–49

Reed LY, Meunch M (1983) A simple method for estimating fifty percent end points. Am J Hyg 27: 493–497

Schlieper P, De Robertis E (1977) Triton X-100 as a channel forming substance in artificial lipid bilayer membrane. Arch Biochem Biophys 184: 207–208

Shin ML, Michaels DW, Mayer MM (1979) Membrane damage by a toxin from the sea anemone *Stoichactis helianthus*. Effect of membrane lipid composition in a liposome system. Biochim Biophys Acta 555: 79–88

Spira ME, Klein M, Kochner B, Yarom Y, Castel M (1976) Ultrastructural changes accompanying the disturbances of neuromuscular transmission caused by *Pardachirus* toxin. Neuroscience 1: 117–124

Sund RB, Olsen G 1981) Net sodium and glucose transport in the jejunam, ileum and colon of anasthetized rats in response to intraluminal theophylline and anionic surfactants. Acta Pharmacol Toxicol 49: 65–71

Van Oss CJ, Absolom DR, Newmann AW, Zingg W (1981) Determination of the surface tension of proteins. Surface tension of native serum protein in agueous media Biochim Biophys Acta 670: 64–73

Varanda W, Finkelstein A (1980) Ion and nonelectrolyte permeability properties of channel formed in planar lipid bilayer membranes by the cytotoxic toxin from the sea anemone *Stoichactis helianthus*. J Membr Biol 55: 203–211

Zlotkin E, Barenholz Y (1983) Studies on the mode of action of the ichthyotoxin pardaxin. In: Zahuranec BJ (ed) Shark repellents from the sea: New perspectives. American Association for the Advancement of Science. Washington DC, p 157

Behavioral Bioassays in Laboratory and Field

S.H. GRUBER[1], E. ZLOTKIN[2], and D.R. NELSON[3]

1 Introduction

The large rapaceous sharks are among the few wild animals that still pose a threat to the safety of humans. Any shark that attains a length of 2 m is potentially dangerous, but the majority of recorded shark attacks have been attributed to less than 50 species. Baldridge (1974) analysed approximately 1700 shark attacks and found that most were confined to five families: The Lamnidae, Carcharhinidae, Orectolobidae, Odontaspidae, and Sphyrnidae. The same groups were implicated in South African attacks (Wallett 1978). Actually, most of the sharks that grow to large size are included in these five families. However, within the five only a few species such as the white *(Carcharodon charcharias)*, tiger *(Galeocerdo cuvieri)*, bull *(Carcharhinus leucas)*, and great hammerhead *(sphyrna mokarran)* can be said to attack humans with any regularity.

Baldridge (1974) reported that only about 100 shark attacks occur annually, but suggested that shark incidents would increase as more people entered the sea. Even so, considering the millions of bathers and divers exposed to sharks each year, shark attack must be considered a rare and very unusual phenomenon. Why then do we expend valuable research time and scarce funds on the search for a truly effective shark repellent?

We suggest at least three reasons: (1) public fear created by a shark attack is often amplified by the media which in turn can affect the wider population by bringing activities such as recreational swimming to a halt. (2) Under certain conditions such as aviation disasters at sea the probability of shark attack can increase from near zero to almost certain. War-time accounts of shark attacks on shipwrecked sailors attest to this and such accounts can dramatically affect the morale of an entire fleet (Zahuranec and Baldridge 1983).

Finally, and of direct importance for our research, sharks attack sensitive oceanographic equipment with some regularity. For exeample, monitoring the acoustic envi-

1 Division of Biology and Living Resources, Rosensteil School of Marine and Atmospheric Science, University of Miami, Miami, FL 33149, USA

2 Department of Zoology, Institute of Life Sciences, The Hebrew University of Jerusalem, Jerusalem, Isr

3 Department of Biology, California State University at Long Beach, Long Beach, CA 90840, USA

Toxins, Drugs, and Pollutants in Marine Animals
(ed. by Bolis et al.)

ronment in the sea requires the deployment of vulnerable hydrophones which may produce electrical and acoustical fields attractive to sharks (Nelson and Gruber 1963; Kalmijn 1978). The result is that sharks can and do destroy acoustic hydrophone arrays. Not only is the equipment lost, but an oceanographie cruise might be cancelled, and if the oceanographic application were in the military realm, a nation's security might be jeopardized. So, for at least these reasons, a modest research effort to continue the search for shark repellents has been funded by the United States Office of Naval Research and the Israeli-US Binational Science Foundation.

The United States Navy has supported research on shark repellency for about 40 years. (for details, see Zahuranec 1983 this Vol.). The original investigations done under wartime exigency were successful to the extent that a chemical repellent was developed and deployed throughout the fleet. However, the experiments upon which the repellent was based were flawed and eventually the compound known as US Navy Shark Chaser had to be withdrawn (Zahuranec 1975).

After World War II research continued and although literally hundreds of substances were tested on large sharks in open pens (Baldridge 1969 a, b), none was found to have the desired quick-repellent action. Even powerful toxins which eventually killed the shark from brief exposure did not produce the desired repellent behavior (Nelson 1983).

Then, in 1972 and 1973 Clark (1983) carried out a series of experiments which strongly suggested that the Moses sole *(Pardachirus marmoratus)* produces a secretion which protects it from shark attack both under laboratory and field conditions. From these studies followed a series of experiments on the sole's secretion (pardaxin = PXN) culminating in a 1981 symposium on shark repellents from the sea (Zahuranec 1983). Three possible lines of future research emerged from the symposium. First, Primor et al. (1983) and Moran et al. (1983 this Vol.) recommended continued physiological study of PXN so that target organ and mode of action could be unambigrously specified. During the symposium, Primor favored the theory that gills were the target organs and "... production of a net sodium influx is the mechanism of PXN induced behavioral responses ..." (p. 181).

On the other hand, Zlotkin and Barenholz (1983), noting the surfactant and amphipathic character of PXN, attributed the toxin's effects to a more general interaction whith membranal phospholipid. They predicted that, "... in the search for shark repellents it may be worthwile to pay attention to surface-active substances ..." (p. 169). Finally, several other authors called for a third line of investigation: The search for and continued testing of a wide variety of naturally occurring toxic compounds.

The single common element in all three lines of research was the study of shark behavior under laboratory and field conditions. Two of us had been studying shark behavior since the early 1960's (i.e., Nelson and Gruber 1963). Nelson has been more interested in field behavior while Gruber has concentrated on the analysis of behavior under controlled conditions.

For the remainder of the paper, we will describe the development by Gruber and Nelson of laboratory and open-sea bioassays for testing Zlotkin's hypothesis that surfactants repel sharks. Through a series of coordinated studies, using these biosassays we will present evidence that certain synthetic and natural detergents repel sharks under a variety of conditions.

2 Materials and Methods

2.1 Animals

The laboratory and field bioassays employed three species, one telelost and two elasmobranchs as test subjects.

1. The goldspot killifish *(Floridicthyes carpio)*, a teleost, is a common littoral inhabitant of the muddy shoreline of Biscayne bay near the Rosensteil School on Virginia Key (Miami), Florida. Several hundred specimens for study were collected by beach seine and maintained in 600 liter marine aquarium prior to testing. The test subjects were 32 to 45 mm (x = 37,6 mm) in total length (TL) and 0.53 to 1.51 g in weight. These fish were carefully identified using Eddy (1969). We chose the goldspot killifish because of its hardiness, availability and ease of capture.

2. The lemon shark *(Negaprion brevirostris)* was the test subject in all of the laboratory bioassays. This shark is found in nearshore waters from New Jersey to Brasil, possibly in the eastern Pacific and tropical eastern Atlantic (Compagno 1978). While the lemon shark attains a length of 300 cm, the approxomately 50 specimins we tested in the laboratory ranged from 61–85 cm TL. Details of maintenance can be found in Gruber (1980). We chose the lemon shark because it is common, easily captured, can be rapidly and reliably trained and subjected to experimental manipulations, adapts well to captivity and is known to be a dangerous shark. One of us (SHG) has worked with this shark for two decades.

3. The blue shark *(Prionace glauca)* was the test subject for the field studies. It is one of the most common and is the most wide-ranging of all sharks. It is found in the epipelagic zone of all tropical to cool-temperate seas (Compagno 1978). The blue shark is the most commonly encountered shark in the surface waters of Los Angeles, California where the tests were carried out, so in one sense, the blue shark chose us. Under ordinary conditions blue sharks can be attracted to a boat with ground fish (chum). Once at the boat, they usually swim slowly, remain in the vicinity and will take a bait. Thus, they are excellent subjects for field tests. Nevertheless, blue sharks are dangerous and have bitten human beings.

2.2 Materials and Compounds Assayed

There were two general types of substances tested: Natural and synthetic surfactants. Since at the start we knew that PMC the crude, lyopholysed extract ot the toxic secretion of *Pardachirus marmoratus*, would repel sharks, we used the effective dose level of this substance as a reference against which to compare effectiveness of other substances. PMC was prepared by collecting and lyophilizing the toxin from a living moses sole, and storing the dried, white powder in sealed vials, in darkness at about 20° C. In general, test substances were obtained in the form of water-soluble, dry powders. The materials were reagent grade and mostly obtained from Sigma Co (USA). Just prior to a test, a stock solution of the required material(s) was reconstituted to various concentrations by mixture with distilled water. Table 1 lists most of surfactants tested. Five unlyzed derivatives of lauric acid (lauryl amine, lauryl alcohol, lauryl bromide, sodium laurate and the ethyl ester of lauric acid - Sigma Co) were tested on sharks and found to be inactive. However, unlike the other materials, they were all insoluble in sea water.

In a seperate series of experiments we tested a second toxic fraction which had been isolated from the defensive secretion to the Japanese sole *P. pavonius* by Tachibana et al. (in press and this Vol.). The isolation by Tachibana of a low molecular, lipid-soluble substance suggested to us that PMC might contain the same sort of fraction. Accordingly, we treated a sample of PMC following the method of Tachibana et al. (in press). 1.05 g of PMC were dissolved in 15 ml of 0.1N NaOH and 150 ml Me_2CO was added to the solution. The resultant mixture was centrifuged at 15.000 g for 10 min. The supernatent was evaporated in a rotovaporator and the final product dissolved three times in a mixture of 15 ml 0.1N NaOH and 15 ml ethylacetate. The organic phases were collected with a separatory funnel and dried to a solid white crystalline substance. Of the original 1.05 g 66% was recovered as 463 mg of protein precipitate and 195.5 mg of an organic extract.

2.3 Bioassays

The laboratory tests employed three bioassays extensively described elsewhere (Gruber 1981, 1983 a, b; Gruber and Zlotkin 1982; Zlotkin et al., 1983 and Zlotkin and Gruber, in press).

1. The response measure used with the *killfish bioassay* was physiological death as indicated by cessation of gill movements. In all tests, an LD_{50} was established by the method of Reed and Meunch (1938). The bioassay was conducted by placing three randomly chosen killifish in a container with 150 ml of filtered sea water. That exposure to various concentrations of test substances which killed one or more of the test fish was recorded. In preliminary assays, substances were added to individual containers in log steps to bring the final concentration from 1 to 1000 mg ml^{-1}. In the critical tests, the final concentrations was doubled in an ascending series between the appropriate log concentrations. The exposure time was usually 8 h. Further details are given in Section 2.4.

2. The response measure used in the *feeding bioassay* was whether a lemon shark, briefly exposed to a test substance continued to feed. In this bioassay a dose which was effective in inhibiting 50% of the tested sharks from feeding (i.e., ED_{50}) was established. The bioassay was conducted in a 12.000 l flowing sea water aquarium of 6 X 2 X 0.5 m. The flow rate was 80 l min^{-1}. Filtered sea water entered one end and drained from the far end of the tank. Prior to a test session, sharks were deprived of food for 48 h. Then a bait was prepared by attaching a 25 cc plastic syringe to a whole blue runner *(Carynx chrysos)*. A polyethylene tube attached to the syringe nozzle protruded from the mouth of the bait. Once offered, the sharks immediately attacked: The bait was then manipulated by the experimentor so that the shark grasped the fish's head. The experimentor immediately released 4 ml of a test substance of known concentration into the shark's mouth and scored the result of this brief exposure (Fig. 1).

We observed graded responses from no effect (i.e., the bait's head was removed and ingested by the shark) to a strong repellent effect i.e., the shark was obviously distressed and disoriented. However the only behavior scored was whether or not the shark fed. Typically, concentration of the test substance was doubled on successive trials. Sharks in the test colony were given only a single exposure on any day. Tests were given twice per week.

3. The third behavioral bioassay was based upon termination of a trance – like state known as *tonic immobility* (TI). An ED_{50} for substances terminating TI was established. Tests were conducted in a 20 l. container through which filtered sea water flowed at a rate of 25 l min^{-1}. A shark was placed in TI by inverting and restraining it underwater in a V-shaped acrylic trough for about 15 seconds. It then lapsed into TI which was signaled by general relaxation, deep rhythmic breathing and occasional tremors. Preliminary study (Gruber and Watsky unpubl.) with lemon sharks given massed trials of 12 inductions per day demonstrated that the subjects could be expected to remain in TI for more than 10 min before spontaneously recovering (Fig. 2).

A test consisted of placing a shark in TI, allowing it to adapt for at least 3 min and then giving "control trials" by injection of sea water into the shark's mouth five times. If the shark remained under TI, a test trial consisting of injecting a known amount and concentration of test substance into the shark's buccal cavity was carried out. The injection was done by hand using a 5 cc syringe to which 3 mm diameter polyethylene tube was attached. The syringe contents were emptied in less than 1 s and emptying was timed to coincide with the start of an inhalation.

Fig. 1. Feeding bioassay: Photograph of a 70 cm TL lemon shark as it attacks the bait and grasps the head in its mouth. Simultaneously, the experimentor releases a measured quantiy of test substance from a syringe, attached to the bait, into the shark's mouth. The result of the trial depends upon the behavior of the shark immediately after this moment

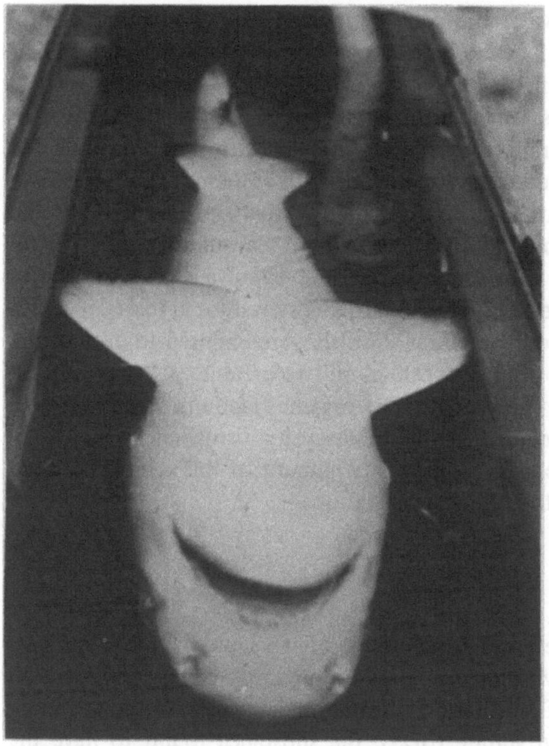

Fig. 2. Tonic immobility assay: Photograph of an 85 cm TL lemon shark under TI in the 20-l test chamber. Induction of TI merely requires the experimentor to restrain the shark in this inverted position for about 30 s. Once in TI the shark will remain immobile and unresponsive for at least 10 min. Known quantities of test substances can either be released into the chamber or into the shark's mouth as it respires. Trial's outcome depends on whether the substance stimulates the shark to right itself

Several qualitative responses were recorded but the trial was counted as positive only it the shark righted itself. A preliminary of tests were given to roughly establish the effective concentration by injecting a single dose of 1, 5, 10, 50 or 100 mg ml^{-1} into the shark's buccal cavity. In the final test, an ascending method of limits was used whereby the shark was exposed to increasing amounts of test material over a period of several trials by doubling the concentration on subsequent tests. The threshold was arbitrarily chosen as the lowest concentration of test substance which terminated TI in 50% of the test subjetcts.

In a later study this procedure was modified by fitting the testing tank with a recirculating pump of 1200 l min^{-1} capacity. The pump allowed for rapid even mixture of a test substance. Trials consisted of releasing a known amount and concentration of test material from a syringe into the pump and observing the behavior of the shark. Again tests were scored as positive if the shark righted itself. Thresholds were calculated as described above.

2.4 The Pavoninin Study

Five substances were assayed in this study: (1) PMC; (2) protein precipitate of *P. pavonius*, a whiter powder presumably containing a protein similar to pardaxin; (3) a mixture of pavoninins (see Tachibana, this Vol.); (4) protein precipitate of *P. mar-*

moratus, 463 mg of a white powder probably containing pardaxin and (5) the organic extract of PMC, 195.5 mg of a white crystal containing a substance very similar to pavoninin (Tachibana, pers. commun).

Because so little pavoninin was available for testing we concentrated on the killifish lethality assay. Tests were arranged by placing three killifish, randomly selected from a population of several hundred, in a container with 150 ml of filtered sea water. Prior controls established that three killifish could survive more than 48 h in the 150 ml of sea water. We also exposed killifish to 150 ml of sea water to which 600 μl of ethyl alcohol was added. (600 μl is times the amount of solvent used to dissolve the organic fractions in any single test). These killifish survived for at least 18 h with no apparent distress. Different groups of three killifish were exposed to each of the five substances at concentrations from 5 to 500 $\mu l\ ml^{-1}$ (see Table 2). Time to death was indicated by cessation of gill movements. The organic fractions of both *P. marmoratus* and *P. pavonius* were prepared for the bioassay by dissolving 18 mg of the crystals in 600 μl of ethyl alcohol. The appropriate amount of this solution was added to 150 ml sea water to make the final concentration.

2.5 The Field Study

Experiments are underway to determine the effects of repellent chemicals on free-ranging sharks in the natural environment. Trials are presently being conducted on pelagic sharks in the waters off southern California, and involve surfactant compounds including sodium dodecyl sulfate (SDS), the substance found to have the greatest effectiveness in the laboratory assays. In addition to providing basis concentration thresholds for repellency for various substances, a second objective of the field work is to study the repellency behavior itself, i.e., orientation mechanisms and other behavioral phenomena such as habituation and learning. The primary study site in California is the San Pedro Channel between Long Beach and Santa Catalina Island. As mentioned, sharks are attracted to be boat and those arriving up a baitodor corridor are usually quite fearless and will approach the boat closely, circling for a prolonged time. Exposures consist of control trials with sea water only, and test trials of sea water solutions of the chemicals.

The following test methods were performed:

1. Delivery to oral cavity. In one version, the test solution is gravity-fed from a resevoir in the boat down a flexible plastic tube which terminates inside a bait fish. Opening a valve on the tube initiates the flow of chemical into the shark's mouth after it has grasped the fish. Another version utilizes a measured amount of chemical in a latex "balloon" packet tied to the bait fish (Fig. 3). When the shark bites the bait, the teeth rupture the packet, releasing the chemical inside the mouth.

2. Squirt application. A diver in a protective shark cage applies a measured amount of solution to the circling shark using an extended bulb-type syringe. Unlike the above procedure, this method does not involve the shark ingesting the bait at the moment of application of the chemical. It may also allow determination of the target organs for repellency, since the material can be applied specifically to various locations on the shark's head, i.e., nostrils, eyes, mouth, and gills.

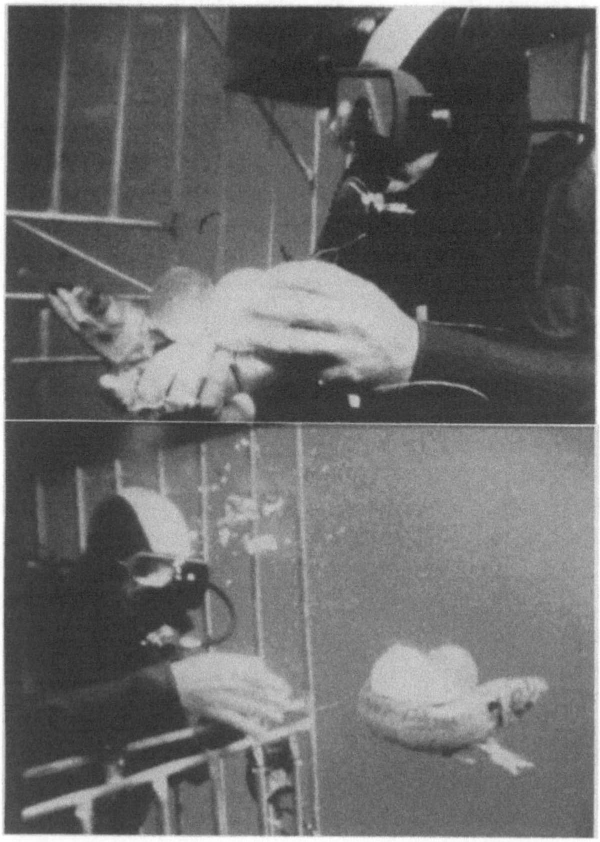

Fig. 3. Diver in shark cage preparing bait fish with test repellent packets (latex ballons containing SDS solution) which rupture when bitten by sharks. (From 16 mm cinema frames by H. Hall, courtesy of R. Johnson)

3. Delivery into odor corridor. At a certain point in a shark's approach up an odor corridor, a measured quantity (or rate) of test solution may be added to the outgoing attractant. Observers will then note if the shark is turned ayway, and at what distance down the corridor.

3 Results

The lavoratory results reported here were obtained in four test sessions between 1981 and 1983. As noted the primary experimental question was whether strong surfactants in general would repel sharks. Another objective was to determine the effectiveness of pavoninin (see Tachibana this Vol.). The field results were obtained during several trips to sea in the summer of 1983.

Table 1. Bioassays of synthetic and natural surfactants using the goldspot killifish. *(Floridicthyes carpio)* and lemon shark *(Negaprion brevirostris)*

Substance Generic and Commercial Name	Bioassays		
	Shark Killifish lethality LD_{50} (μg ml^{-1})	Shark tonic immobility ED_{50} (mg ml^{-1})	Feeding assay range (mg ml^{-1}
A. Sodium lauryl sulfate (SDS)	3.0	0.45	0.2–2.0
B. Lithium lauryl sulfate (LDS)	6.0	0.62	0.2–2.0
C. lyophilysed crude secretion of *P. marmoratus* (PMC, non-synthetic)	16.0	0.66	0.8–3.0
D. Lauryl trimethyl ammonium	60.0	8.0	3.0–8.0
E. Polyethoxylated octyl phenol. (Triton-X-100)	36.0	10.0	6.0–8.0
F. Sodium cholate[a] (nonsynthetic bile salt)	> 100	8.1	8.0–10.0
G. Ethoxylated-20-sorbitan monolaurate (tween 20)	> 100	10.0	10.0–20.0
H. Saponin (nonsynthetic mixture of steroidic glycosides)	> 10	100	~50
I. Ethoxylated-100 sterate[b] (Myrj 59)	1000	> 50	~50
J. Ethoxylated-23-lauryl alcohol[c] (Brij 35)	>10	> 50	>50
K. Deca glycerol monooleate (10•G•1•0)	1000	>100	> 50
L. Zwitterionic derivative of cholic acid (nonsynthetic "Chaps")	>100	> 10	Not tested

[a] Critical lethality test not done because of high concentration of substances F-L needed to cause death.

[b] Only a few TI trials were done with substances H-L because of high concentrations needed to terminate TI.

[c] Only a few feeding trials were done with substances J, K and L because of high concentrations needed to inhibit feeding.

3.1 Laboratory Studies

Table 1 summarizes results of the surfactant bioassays and includes all the tests except those with the insoluble surfactants, protein precipitates and the pavoninins. Inspection of table 1 shows that it was possible to rank the most effective substances (A–F) because of the close correspondence between the three assays. The substances are ranked according to the feeding assay. However only substances E and F could be interchanged. Otherwise, the correspondence between all three bioassays was good. Substances A and B were the most effective in all assays. In the killifish bioassay substances I, K and L may be considered nontoxic, F, G, H and J midly toxic and A, B, C, D and E quite toxic. The 6 h (LD_{50} for SDS is 3 ppm by weight.

In the shark-feeding test, substances H through K in ceoncentrations below 40 mg ml^{-1} did not deter hungry lemon sharks from feeding, whereas substances D through G were moderately effective at inhibiting feeding but did not cause obvious distress to the sharks. Substances A, B, and C strongly repelled sharks. For example, substance C evoked a strong negative response at 10 mg ml^{-1} and a moderate one at 2 mg ml^{-1}. Substance A clearly and reliably replled sharks. Even at 0.8 mg ml^{-1} five of seven sharks tested with substance A were instantly repelled and showed signs of distress as they sped away from the bait. Higher concentrations of the effective substances on the order of 3–5 mg ml^{-1} produced obvious signs of distress such as disorientation, muscular tremors, and chances in coloration which we have previously associated with respiratory insufficiency (Gruber 1980). One shark inadvertantly tested twice on one day with 5 mg ml^{-1} SDS became immobile 5 min after the 2nd trial and ceased breathing after 20 min. The shark was revived by directing of oxygenated water through its gills for 2 h. Clearly SDS and LDS were more effective at repelling sharks in these laboratory studies than the natural PMC. However, all the other surfactants were either mildly repellent or ineffective.

Table 2. Bioassay of pavoninin and related substances using the goldspotk killifish *(Floridicthyes carpio)*

Substance	[a] Concentration of substance used in killifish lethality bioassay ($\mu g\ ml^{-1}$)					
	5	10	20	50	100	500
1. Crude secretion of *Pardachirus marmoratus* (PMC)	0/3[b]	0/3	3/3[c]	3/3	–	–
2. Organic extract of *P. Pavonius* (Pavoninins)	0/3	0/3	3/3	3/3	–	–
3. Organic extract of *P. Marmoratus*	0/3	3/3	3/3	3/3	–	–
4. Protein precipitate of *P. Pavonius*	–	–	–	0/3	0/3	0/3
5. Protein precipitate of *P. Marmoratus*	0/3	0/3	0/3	2/3	3/3	–

[a] Assay lasted approximately 7.5 n.

[b] 0/3 refers to 0 fish of the 3 died during the exposure at that concentration, i.e., all fish survived. In contrast 3/3 means that 3 of fish died.

[c] The LD_{50} which lies between 10 and 20 $\mu g\ ml^{-1}$ PMC corresponds closely to the LD_{50} of PMC in Table 1

The tonic immobility tests gave essentially the same results. However, in the modified assay using the pump dispersion method trials were performed in succession with five different sharks. Results showed that the effective concentration of SDS needed to terminate TI was 35.6 ± 3.4 SD $\mu g\ ml^{-1}$. In otherwords, the shark was 12 times more sensitive to an evenly dispersed concentration of SDS bathing its entire body than to mouth presentations. The TI threshold level for SDS was thus 35 ppm.

2. The pavoninin studies: Results of the killifish bioassay are shown in Table 2. Four of the five substances were ichthyotoxic. The protein precipitate of *P. pavonius* was inactive even at 500 $\mu g\ ml^{-1}$. The most toxic fraction was the organic extract of the moses sole which was about 2 times more active than pavoninin and about 5 times more active than the water soluble protein of *P. marmoratus*. PMC possessed the same ichthyotoxicity in these tests as in the previous ones. (see Table 1). By referring the LD_{50} values of PMC to 1 g, we estimate that the organic fraction comprises approximately 60% of the toxicity, while the water-soluble protein provides 40%.

Because of the limited quantitiy of test substances systematic tests with sharks were not done. However 12 TI trials with a single 65 cm male were done. In successive trials, the shark was exposed to 5 ml mouth applications of the four active substances at 4 mg ml^{-1}. The shark reacted only in trials with PMC. Thus pavonin and the extracts of PMC were ineffective at terminating TI. A feeding trial with pavoninin at 4 mg ml^{-1} inhibited the feeding of a single lemon shark. Obviously these trials are inadequate to draw conclusions; but they suggest that pavoninin may not be as effective as PMC in altering the behavior of sharks.

3.2 Field Studies

Field trials to date have been on blue sharks, and have utilized the surfactant SDS (specifically the commercial grade product known as sodium lauryl sulfate, which is about 98% pure and available in liquid form as a 30% aqueous solution). Initial tests by Robert R. Johnson and Jeffrey Landesman of the California State University, Long Beach shark-research group confirmed that SDS in relatively concentrated form has a dramatic repellent effect when delivered into the mouths of sharks feeding on bait. As shown in Fig. 4, about 100 ml of a 15% solution causes immediate rejection of the bait, nictitating membrane closure, and a rapid withdrawal with the mouth held wide open. A 1% solution (tube-delivered) also repelled, but less dramatically with response latencies of several seconds. These are preliminary values, undoubtedly well above response thresholds. In underwater tests from a shark cage, diver R.R Johnson, Cabrillo State museum, Los Angeles squirtied 3% SDS into the faces of blue sharks and obtained moderate repellent effects, but the exact quantity of material reaching the sharks was not determined.

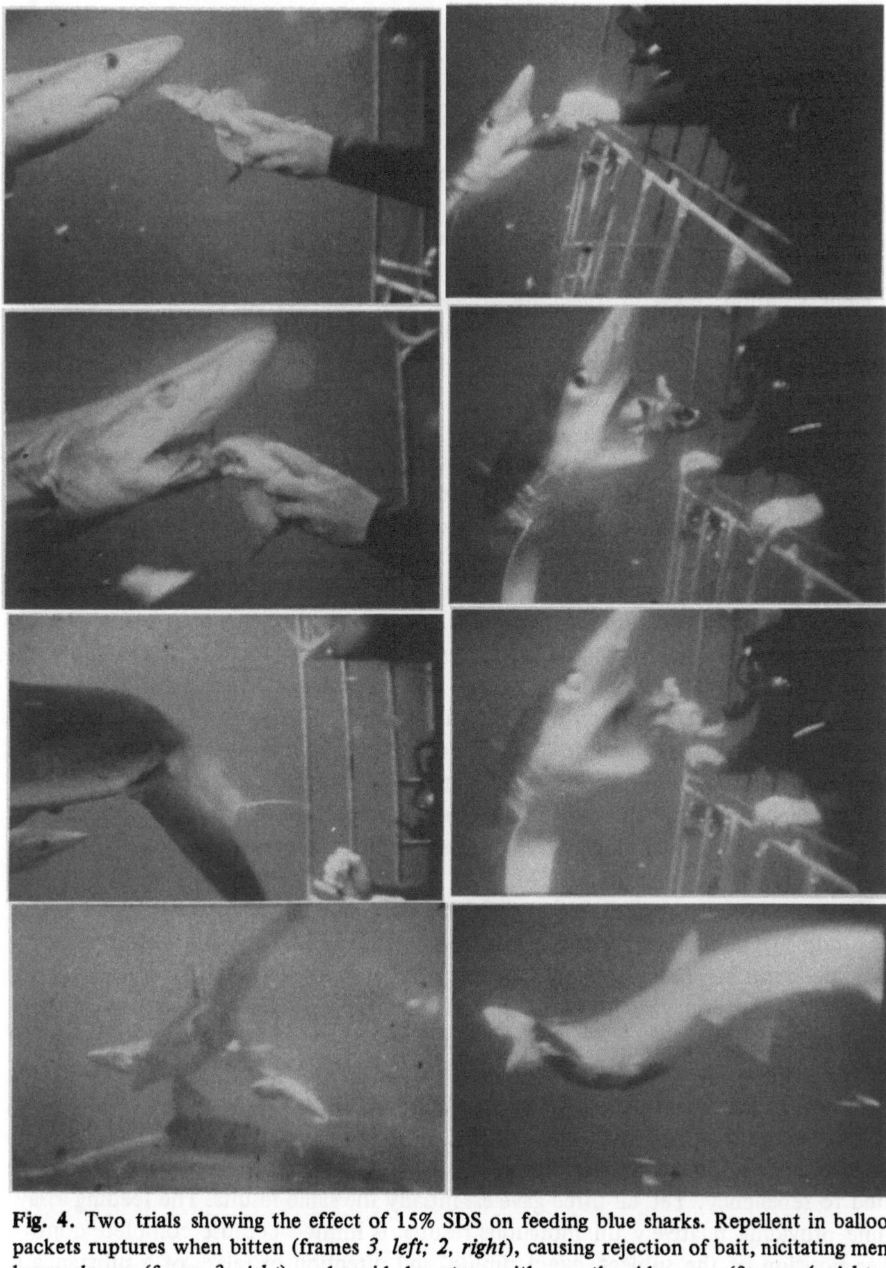

Fig. 4. Two trials showing the effect of 15% SDS on feeding blue sharks. Repellent in balloon packets ruptures when bitten (frames *3, left; 2, right*), causing rejection of bait, nicitating membrane closure (frame *3, right*), and rapid departure with mouth wide open (frames *4, right; 4, left*). (From 16 mm cinema by H. Hall, courtesy of R. Johnson and Don Meier Productions)

4 Discussion

Shark attacks may be deterred by physical devices such as barriers or by altering the attractive stimulus qualilties of an object so as to render it aversive. Our experimental approach to chemical repellents is based on the molecular characteristics and pharmacology of the skin scretion of the moses sole. Several lines of evidence detailed in Zlotkin and Barenholtz (1983) and Zlotkin and Gruber (in press) led to the hypothesis that the shark-repellent properties of pardaxin could be attributed to its surfactant characteristics, i,e., PXN shows foaming and drop volume reduction in aqueous solution, lyses artificial liposomes, etc. (see Moran et al., this Vol. further details). It follows from this hypothesis that synthetic and other easily obtained surfactants may repel sharks. This is important because pardaxin is a rare and labile substance and as such impractical for use as a shark repellent.

The purpose of this study was twofold: (1) to develop rapid and reliable bioassays for use in the laboratory and open-ocean to screen potential shark repellents and (2) to use the assays to test Zlotkin's hypothesis that synthetic and other surfactants may repel sharks. The finding of any compound that repels sharks is not trivial. Over the past 40 years a wide variety of chemical agents such as fish poisons (rotenone), irritants (chlorine), systemic poisons (cyanide), immobilizers (tricaine) and chemical warfare agents have been tired as shark repellents (Baldridge 1969 a, b; Gilbert 1970). While sharks were killed by some tests, the desired behavioral effect, i.e, rapid withdrawal by sharks did not occur at the low concentrations deemed practical. Surprisingly, most of the test were entirely negative (Nelson 1983). In other words, sharks are generally unresponsive to noxious and even lethal chemicals which would appear to be very irritating to other organisms. Thus to find a common, relatively mild compound which would activate and repel sharks represents a significant departure from past results and leads to a renewed optimism in the search for chemical shark repellents.

We have carried this search forward by developing appropriate test methods for the study of potential repellents in laboratory and field. In the past development of such bioassays has been hampered by the difficulty of keeping sharks in captivity (Myrberg 1976) and working with them in the field. However, recent progress in comparative toxocology, and in behavior and husbandry of elasmobranchs, as well as the study of free-ranging sharks (Gruber 1983 a, b; Gruber and Myrberg 1977); Nelson 1977, 1978), has provided the background to enable us to accomplish the studies reported here.

The laboratory bioassays are diverse and only one, the feeding assay, is directly related to repellency. Yet, all three gave essentially the same results. The feeding assay, while providing relatively unambiguous results, is limited because complete control over position of the subject, exact timing, and direction of relase of a substance is not possible. The shark's motivation is also an uncontrolled variable, and while conditions can be arranged so that motivation to feed is high, once it obtains food, the shark's hunger rapidly declines. The practical results is that only a few trials can be performed each week.

Thus we developed the TI assay, which is more independent of the shark's motivation, but less valid as an indicator of repellency. Results of a preliminary unpublished study demonstrated that lemon sharks do not habituate to TI trials. Rather, when placed under TI sharks can be expected to remain essentially immobile for 37 ± 2.5 SE min with no decrement over time. The TI method, when compared to the feeding assay, consists of a repeatable series of stereotyped behavioral events which are resistant to habituation and have quantifiable starting and end points. The agreement between the three assays and the especially close quantitative and qualitative relation between the feeding and TI assay (Table 1) suggests that tonic immobility may be employed for the rapid screening of potential repellents. The easy availability of such a simple and rapid test means that large numbers of compounds can be quickly screened and the most promising candidates set aside for further testing.

The results in Table 1 show that Zlotkin and Barenholtz's (1983) prediction that synthetic surfactants possess shark-repellent properties was, in principle verified. Both Sodium and lithiumn lauryl sulfate were 5 times more ichthyotoxic and at least 4 times more repellent to lemon sharks than PMC. The field trials also confirmed that sodium lauryl sulfate is, under relatively natural conditions, a potent shark repellent. Yet the relation between the chemical properties of surfactants and their repellent properties remains to be clarified. In this study we used all types of surfactants: natural, synthetic, cationic, anionic, and nonionic. However only the anionic, lauryl sulfate salts were effective. We must conclude that a general surfactant interaction with membranal phospholipids does *not* account for the repellent effect. Moran et al. (this Vol.) arrived at the same conclusion from a completely different set of results and perspective. They suggest that since both SDS and PXN produce hydrophobic channels in lipid membranes, these channels and the resultant sodium influx might account for the repellent effect. One problem with this explanation is that Triton-X-100, a detergent one order of magnitude less effective than PMC or SDS, also forms channels (Schlieper and DeRobertis 1977).

Another factor mitigating against a general surface active explanation for shark repellency is that SDS is effective as a repellent in sea water solutions well below its critical micelle concentration (Elworthy 1968). Thus, we agree with Moran et al. (this Vol.) that the channel-forming hypothesis and indeed other hypotheses should be tested until the physiological mechanism of shark repellency by surfactants is clarifield.

It is evident from these results that SDS has a considerable degree of shark-repellent action. The question remains as to how effective it would be as a practical repellent for human use, and testing now in progress should provide an answer. Nelson (1983) reviewed the biology of chemical repellency in general, including past efforts to develop a chemical repellent for sharks. He pointed out that the usefulness of a chemical method depends on the particular attack situation involved (i.e., feeding versus agonism, presence or absence of distress stimuli, etc.), whether the person can see the approaching shark, and the duration of protection desired. For example, providing adequate protection for a scuba diver working among sharks is a much different problem than protecting a victim (possibly wounded) of an air/sea disaster.

Because of the generally negative results of past repellent testing, some researchers have been pessimistic about the chemical approach to shark repellency (Baldridge 1974, 1976; Gilbert 1976). This pessimism was based partly on the problem of the relatively large amount of chemical required to protect someone by maintaining a cloud of dissolved repellent at the needed concentration around that person especially in heavy wave action. Unless an extremely potent – and presently unknown – repellent is discovered, the quantitiy of material required becomes excessive. Baldridge (1969 a, b) conducted a mathematical analysis of the dispersive action of water-borne drugs. Combining this with his findings on the speeds of action of these chemicals on sharks, he concluded that the chemical repellency of sharks was completely impractical.

Baldridge's analysis, however, was based on substances acting as toxins or other physiological incapacitants – and these have a considerable latency of action due to the time needed for the material to pass through the gills and be distributed by the blood to the appropriate internal sites of action. What is needed is a true repellent, a substance which acts essentially instantaneously as a strong irritant to external receptors. SDS appears to act in just this way (Gruber and Zlotkin 1982; Moran et al. this Vol.). These use of such a fast-acting, true repellent, therefore, reduces the objections of Baldridge – but there is still the quantitiy problem of the "surrounding-cloud" method of presentation.

Probably the most effective way to utilize existing repellent chemicals is to deliver a concentrated dose, aimed at the shark's head, perhaps even specifically at the most sensitive areas such as the eyes, nostrils and gills.

Several versions of this method were suggested by Nelson (1983), including the (1) squirt gun (underwater "Mace"), a simple plastic squeeze bottle or syringe-type gun, perhaps on a telescoping pole, (2) repellent-tipped shark billy, a short pole used to push away sharks, enhanced at the tip with repellent chemical in a slow-relase matrix and (3) repellent injector (stinger), a pole tipped with a short hypodermic needle/syringe to inject a small quantity of strong irritant to the subdermal free nerve endings much like a bee sting.

We believe that a useful, practical repellent "weapon" can be produced along the lines suggested above, using a repellent no more potent than SDS, which also has the important advantages of being inexpensive and inoffensive to humans. Such a method, of course, would be of most use to persons equipped to see clearly underwater, such as divers. It is somewhat less adaptable to the classic survival situation, in which a victim of an air/sea disaster may be drifting helplessly in a life jacket awaiting rescue. For such situations, a Shark-Screen deterrent (i.e., a man-sized plastic bag with inflateable collar; Johnson 1972; Nelson 1983) may be preferrable to a strictly chemical method, although a chemical device would certainly aid the survivor to discourage inquisitive sharks from closely circling the bag. One suggestion for any marine survival is a diver's face mask to permit clear viewing of approaching sharks.

The major finding of the pavoninin study is that PMC from the Moses sole contains a second active factor, a steroid substance very similar to the pavoninins of the toxic Japanese sole. The steroid toxin of *P. marmoratus* appears to be more ichthyotoxic than that of *P. pavonius.* We were surprised that the water-soluble protein fraction of *P. pavonius* was inactive in lethality tests on the goldspot killifish. Tachibana et al. (in press) suggested that the water-soluble protein might be more active than the pavoninins because it presumably contains the potent toxin, paradin. Indeed they reported lethal activity when tested on the Japanese killifish. Thus our results differend from theirs. We can speculate that precipitation of PXN by acetone is a robust treatment and might reduce the final potency of PXN. However, our calculation of the relative toxicity of the steroid and protein components suggests that the activity of PXN is not greatly reduced by precipitation. For example, comparing the LD_{50} value of PMC to that of both the protein precipitate and the lipid-soluble fraction then taking into account the yield from fractionation we estimate that 66% of the original 1.05 g was recovered and 61%, i.e., most of the expected toxicity was realized. Taken together, the results show that the steroid portion of PMC was 1.5 times more toxic than the protein protion.

The shark tests were less revealing. In the few trials recorded PMC was more active than pavoninin in both TI and feeding trials. Tachibana (this Vol. speculated that pavoninin may be a repellent acting on the shark's olfactory system, while pardaxin might inhibit feeding. Our results did not confirm this conclusion. However, the finding of a second and previously unknown component of the toxin raises important questions about the complete mode of action of the sole's defensive secretion. For example, the a steroid toxin is now described from two species of *Pardachirus* and must be of general evolutionary importance in the defensive strategy of this genus. Added to this is the unknown, but important defensive role of the mucus coating. Clark (1983) showed that removal of the mucus immediatley put the Moses sole at risk and sharks quickly consumed them. Thus, where before we believed that a single factor, pardaxin, accounted for repellency, we now recognize three factors perhaps acting in concert. This raises new hope that a truly effective shark repellent acting in very low concentration may yet be found through study of this unique genus of toxic soles.

Acknowlegements. This study was supported by contracts from the US Office of Naval Research to SHG and DRN and by grant 2983/83 of the IsraelUS Binational Science Foundation to EZ. The authors thank Professor Y. Barenholz and N. Garti from the Hebrew University for their encouragement. Capt. Warren Servatt collected the sharks. We appreciate his extra efforts. Professor Koji Nakanishi kindly provided materials for testing and he and Dr. Y. Naya, through the Suntory Institute for Bioorganic Research supported the visit of SHG to Osaka for consultations about Pavoninin. We are very grateful for their support. Finally, we thank the organizers of the symposium, especially the elegant Jose Zadunaisky for adding a touch of class to our otherwise technical presentation. Ms Karen Seykora typed the manuscript.

References

Baldridge H Jr (1969 a) Kinetics of onset of responses by sharks to waterborne drugs. Bull Mar Sci 19: 880–896

Baldridge H Jr (1969 b) Analytic indication of the impracticability of incapacitating an attacking shark by exposure to waterborne drugs. Mil Med 134: 1450–1453

Baldridge H Jr (1974) Shark attack. Droke House/Hallus, Anderson SC

Baldridge H Jr (1976) A reminder, of the impracticability of chemical shark repellents. In: Seaman W Jr (ed) Sharks and man: A perspective Florida Sea Grant Program Report 10. Gainseville, p 18

Clark E (1983) Shark repellent effect of the Red Sea moses sole. IN: Zahurance B (ed) Shark repellents from the sea. AAAS selected symposium, Washington DC, p 135

Compagno L (1978) Sharks. In: Fischer W (ed) FAO identification sheets for fishery purposes. Western Central Atlantic, fishing area 31, Vol V. FAO, Rome

Eddy S (1969) How to know freshwater fishes, 2 nd edn. WC Brown, Boston

Elworthy P, Florence H, McFarlane C (1968) Solubilization by surface active agents. Chapman and Hall, New York
Gilbert P (1970) Studies on the anatomy, physiology and behavior of sharks. Unpublished final report to the US Office of Naval Research, Arlington
Gilbert P (1976) An evalutation of some chemical, biological, and physical agents tested for their effectiveness as shark deterrents. In: Seaman W Jr (ed) Sharks and man: a perspective. Florida Sea Grant Program Report 10, Gainseville, p 19
Gruber S (1980) Keeping sharks in captivity. J Aquaric 1: 6–14
Gruber S (1981) Shark repellents: persepectivies for the future. Oceanus 24: 72–76
Gruber S (1983 a) Shark repellents: protocols for a behavioral bioassay. In: Zahuranec B (ed) Shark repellents from the sea. AAAS selected symposium, Washington DC, p 91
Gruber S (1983 b) A second active factor from the toxic secretion of the Red Sea sole *Pardachirus marmoratus*. Technical report UM/RSMAS 83014 to ONR code 422CB, Miami
Gruber S, Myrberg A Jr (1977) Approaches to the study of the behavior of sharks. Am Zool 17: 471–486
Gruber S, Zlotkin E (1982) Bioassay of surfactants as shark repellent. Nav Res Rev 34: 18–27
Johnson C (1972) Conuntermeasures to shark attack. In: Pickwell G and Evans W (eds) Handbook of dangerous animals for field personnel. Naval Undersea Center, San Diego, p 123
Kalmijn A (1978) Electric and magnetic sensory world of sharks, skates and rays. In: Hodgson E and Mathewson R (eds) Sensory biology of sharks skates and rays. Office of Naval Research, Arlington, p 507
Myrberg A (1976) The behavior of sharks – a continuing enigma. Nav Res Rev 24: 1–11
Nelson D (1977) on the field study of shark behavior. Am Zool 17: 501–507
Nelson D (1978) Telemetering techniques for the study of free ranging sharks. In: Hodgson E and Mathewson R (eds) Sensory biology of sharks, skates and rays. Office of Naval Research, Arlington, p 419
Nelson D (1983) Shark attack and repellency research: an overview. In Zahuranec B (ed) Shark repellents from the sea. AAAS selected symposium, Washington DC, p 91
Nelson D, Gruber S (1963) Sharks: attraction by low frequency sound. Science (Wash DC) 142: 975–977
Primor N, Zadunaisky J, Tu A (1983) Pardaxin: its action, toxicity and structure. In: Zahuranec B (ed) Shark repellents from the sea. AAAS selected symposium, Washington DC, p 173
Reed L, Meunch H (1938) A simple method for estimating fifty percent endpoints. Am J Hy 27: 493–497
Schlieper P, De Robertis E (1977) Triton X-100 as a channel forming substance in artificial lipid bilayer membrane. Arch Biochem Biophys 184:207–208
Wallett T (1978) Shark attack and treatment of victims in Southern African water. Prunell, Capetown
Zahuranec B (1975) Shark research: present status and future direction. ONR report ACR-208, Arlington
Zahuranec B (1983) Shark repellents from the sea. AAAS selected symposium, Washington DC
Zahuranec B, Baldridge H Jr (1983) Shark research and the United States Navy. In: Zahuranec B (ed) Shark repellents from the sea. AAAS selected symposium, Washington DC, p 1
Zlotkin E, Barenholtz Y (1983) On the membranal action of pardaxin. In: Zahuranec B (ed) Shark repellents from the sea. AAAS selected symposium, Washington DC, p 157
Zlotkin E, Barenholtz Y, Gruber S (1983) Pardaxin versus detergents: a new approach to the development of shark repellents. In: Thompson M, Latif A and Bayoumi A (eds) Proceedings of the international conference on marine sicence in the Red Sea. Bull Inst Oceanogr Fish 9: 419–426

Toxins from Marine Organisms: Studies on Antifouling

G. J. BAKUS and M. KAWAGUCHI[1]

1 Introduction

Field and laboratory studies indicate that toxicity to fishes by benthic marine invertebrates varies inversely with latitude. Usually more than half of the species of exposed sponges, soft corals, gorgonians, holothurians and crinoids are toxic in the tropics (Bakus 1981). Natural toxins play an important role in community structure (Sammarco et al. 1982). Marine toxins are useful in studies on the mechanisms of nerve transmission across synapses (McClure et al. 1980). They show potential in serving as shark repellents (Bakus 1983). They may also be useful as antifouling agents.

Preliminary experiments were conducted in which crude extracts were prepared from a toxic gorgonian and two species of toxic sponges collected from coral reefs off Veracruz, Mexico. These were combined with abietic acid and the solvent dichloromethane and later poured into grooves in dried wood panels. The panels were suspended in sea water for one month (March 20 – April 20, 1982), the crude extracts gradually leaching into the water. The results indicated a general inhibition of settling of certain fouling algae (*Enteromorpha* sp.), and of polychaete worms (*Spirorbis* sp.) by a gorgonian. There was an apparent enhancement of settling *Spirorbis* sp. (by a sponge) and of the bryozoan *Bugula neritina*, the inhibition of *Enteromorpha* sp. presumably providing space for these invertebrate larvae to establish themselves (Bakus et al. 1983). These results led to the question of whether extracts from marine invertebrates in southern California would also cause strong inhibitory effects on the settling of marine organisms in Los Angeles Harbor.

2 Methods and Materials

Test animals were collected from Big Fisherman's Cove, Santa Catalina Island, California on October 29, 1982 and frozen. Crude extracts were prepared from four species of algae and five species of invertebrates, using a mortar and pestle and the solvent ethanol. Antifouling experiments were conducted at the University of Southern

1 Department of Biological Sciences, University of Southern California, Los Angeles, CA 90089–0743, USA

Toxins, Drugs, and Pollutants in Marine Animals
(ed. by Bolis et al.)

California Fish Harbor Laboratory. Plywood panels (16 cm × 16 cm × 1.3 cm) were seasoned in filtered sea water for 5 days. A mixture of 15 ml crude extract, 35 ml dichloromethane, and abietic acid was poured into 24 offset holes per panel. The holes measured 6 mm in diameter and 10 mm deep. The solvent was allowed to evaporate for 48 h, leaving the crude extract suspended in abietic acid, which would slowly dissolve into the sea water. Six panels were arranged in a wooden frame for each test species and two additional sets of six panels served as controls. The panels were submerged 1.5 m apart at a depth of 1.0 m during November 1982, then pre-

Table 1. Effects of crude extracts on common settling fouling organisms in Los Angeles Harbor[a]

	Fouling Organism		
Crude Extract Species	*Ciona intestinalis* (ascidian)	*Bugula neritina* (bryozoan)	*Spirorbis sp.* (polychaete worm)
Controls	40	37	24
Dictyopteris zonarioides (brown alga)	50[b]	32	23
Pelvetia fastigiata (brown alga)	25[b]	40	24
Pterocladia sp. (red alga)	35	29	26
Gracilaria sp. (red alga)	60[b]	33	26
Leucetta losangelensis (sponge)	46	26	27
Axinella mexicanus (sponge)	45	55[b]	22
Lophogorgia chilensis (gorgionian)	42	33	27
Muricea californica (gorgonian)	40	44	18
Parastichopus parvimensis (holothurian)	35	50	20

[a] Animal data (12 panel sides/species; total of 24 control panel sides) are mean densities/225 cm^2. Mann-Whitney U test.

[b] = significant ($P = \leqslant 0.05$).

served in 70% ethanol. Fouling organisms more than 1 cm from the edge of the frame were counted to avoid edge effects and the data were analyzed using the Mann-Whitney U test.

3 Results

The results of the fouling experiments are presented in Table 1.

4 Discussion

The fouling species counted in these experiments were the same as those used for tests with tropical species, as described in the introduction. The common green alga, *Enteromorpha* sp., occurred in small numbers because the edges of the wood frames holding the panels were oriented toward the sun and perhaps also because of seasonal differences in their abundance. They were ignored due to their very low standing crops, in contrast to their strong effects on settling larvae in the previous study.

The only case of antifouling was that of *Pelvetia fastigiata* which adversely affected the larval settlement of the ascidian *Ciona intestinalis. Dictyopteris zonaroides* and *Gracilaria* sp. enhanced the settling of *Ciona intestinalis* and the sponge *Axinella mexicanus* enhanced the settling of the bryozoan *Bugula neritina.* DeLara (1972) showed that the algae *Dictyopteris undulata, Dictyota flabellata,* and *Laurencia pacifica* from Santa Catalina Island, California are toxic to fish and Fenical et al. (1972) and Faulkner (1977) indicated that *Dictyopteris zonaroides* and *Laurencia* spp. contain sesquiterpenes.

These preliminary experiments suggest that tropical species may be more active in producing antifouling compounds than are temperate latitude species. Biochemists have discovered considerably more toxins, antibiotics and a variety of unique compounds in tropical than in temperate species in general (William Fenical, Phillip Crews, and Kenneth Rinehart, Jr., pers. commun). These results are in contrast with those of Bergquist and Bedford (1978), who indicated that sponge antibacterial activity is more prevalent in temperate (87% of species tested) than in tropical latitudes (58% of species tested). What can be concluded from these and other reports? First, differing results may be a product of the techniques involved. For example, standard laboratory tests for antibiosis vs. field experiments on antifouling; the use of alcohol vs. deionized water for crude extracts; the use of dichloromethane vs. alcohol for solvents. There may also be regional or even seasonal differences in the production of classes of compounds irrespective of latitude. The fact that *D. zonaroides* is toxic to fish yet enhances the settlement of *C. intestinalis* indicates the chemical complexity of the marine organisms.

The most efficient method by which we can obtain a better understanding of the role of marine natural products in the environment is for biologists and chemists to conduct joint research programs or to cooperate in other ways. Experiments on the ecological role of natural products need to be conducted in the field as well as in the laboratory or spurious interpretations may result. We also need to consider further uses for marine natural products, for example, such as the use of marine toxins in biodegradable pesticides for agricultural crops.

Aknowledgments. We thank Carol Stepien and Lin Craft for their help with field colletions and identifications. Dr. Bernard C. Abott provided research facilities at the Fish Harbor Laboratory, for which we are grateful. Dr. Richard Brusca kindly reviewed the manuscript.

References

Bakus GJ (1981) Chemical defense mechanisms and fish feeding behavior on the Great Barrier Reef, Australia. Science (Wash DC) 211: 497–499

Bakus GJ (1983) Toxicity in shallow marine waters: potential for developing shark repellents. In: Zahuranec BJ (ed) Shark repellents from the sea: new perspectives. AAAS symposium 83. Westview, Boulder, pp 185–199

Bakus GJ, Evans T, Mading B, Kouros P (1983) The use of natural and synthetic toxins as shark repellents and antifouling agents. Toxicon [Suppl. 3]: 25–27

Bergquist PR, Bedford JJ (1978) The incidence of antibacterial activity in marine Demospongiae; systematic and geographic considerations. Mar Biol (NY) 46: 215–221

DeLara-Isassi G (1972) A preliminary study on the toxic properties of twenty-eight species of marine benthic algae from California and Mexico. MS thesis, Department of Biological Sciencies. University of Southern California, 17 p

Faulkner DJ (1977) Interesting aspects of marine natural products chemistry. Tetrahedron 33: 1421–1443

Fenical W, Sims JJ, Wing RM, Radlick PC (1972) Zonarene, a sesquiterpene from the brown seasweed *Dictyopteris zonarioides.* Phytochemistry (Oxf) 11: 1161–1163

McClure WO, Abbott BC, Baxter D, Hsiao TH, Satin LS, Siger A, Yoshino JE (1980) Leptinotarsin: a presynaptic neurotoxin which stimulates the release of acetylcholine. Proc Nat Acad Sci USA 77: 1219–1223

Sammarco PW, Coll JC, La Barre S, Willis B (1982) Competitive strategies of soft corals (Coelenterata: Octocorallia): allelopathic effects on selected scleractinian corals. Coral Reefs 1(3): 173–178

Venoms and Venom Glands of Marine Molluscs

R. FÄNGE[1]

1 Introduction

The phylum Mollusca, consisting of about 100,000 living species, is divided into seven classes. Venom glands occur in molluscs belonging to the classes Gastropoda and Cephalopoda. Toxins acquired from the environment, such as paralytic shellfish poisons, are found in certain Pelecypoda (bivalves), but also some gastropods have been found to accumulate exogenous toxins. The present article summarizes information on production of venoms in marine molluscs. The field has previously been reviewed a few times: Endean (1972), Baslow (1977), Halstead and Halstead (1978). Scheuer (1973) surveyed the chemistry of compounds isolated from marine organism. The systematic relationship of some of the groups discussed are shown in Table 1. The shells of some types of marine gastropods are illustrated in Fig. 1

2 Gastropoda (gastropods)

Tonnacea. This superfamily includes the families Tonnidae, Cassidae, Cymatiidae, Ficidae and Bursidae. These comprise predatory marine gastropods feeding on various invertebrates. They have remarkably lage salivary glands. Tonnidae (Tun shells). *Tonna (Dolium)* galea (Giant tun shell) is found in the Mediterranean (although rare) and in other subtropical seas. It may grow to the size of a child's head (Fig. 2). The shell is inflated and rather thin. *Tonna* preys on holothurians (Wilson and Gillet 1974).

Cassidae (Helmet or cameo shells). The helmet shells feed upon echinoid and spatangoid sea urchins (Hughes and Hughes 1971). They live in tropical and warm, temperate seas. Some species are very large with thick shells, which are utilized for making cameos. The King helmet shell, *Cassis tuberosa* (West Indies) feeds upon *Diadema antillarum*, a sea urchin famous for its needle-sharp and brittle, probably venomous spines, which can be up to 30 cm long (Randall et al. 1964). When a *Cassis tuberosa* encounters a sea urchin, the latter becomes alarmed and tries to escape. The hunting goes on in slow motion until usually the sea urchin is captured.

1 Department of Zoophysiology, University of Göteborg, Göteborg, Sweden

Toxins, Drugs, and Pollutants in Marine Animals
(ed. by Bolis et al.)

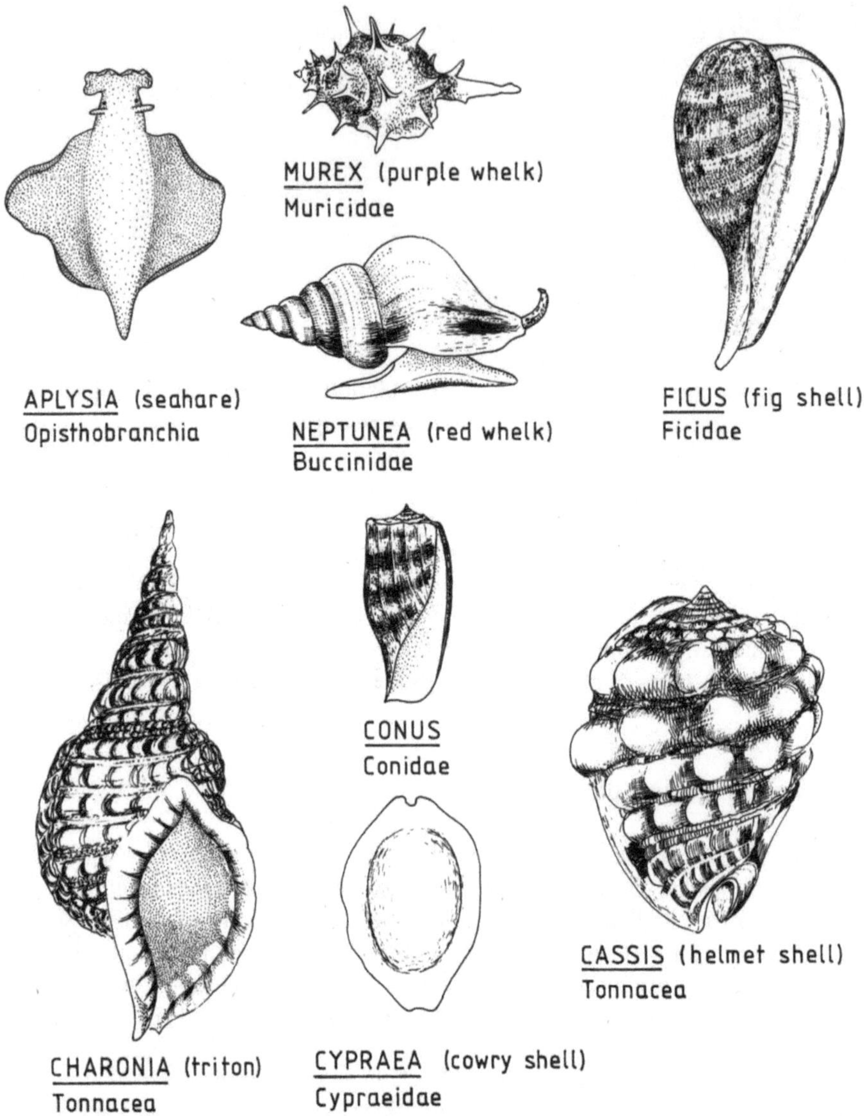

Fig. 1. Types of marine gastropods discussed

By means of its radula apparatus the snail bores a wide hole through the sea urchin's test and ingests its inner organs by means of a long proboscis (Schroeder 1962; Snyder and Snyder 1970). Boring is probably aided by a softening action of the King helmet's acid saliva. *Cassidaria echinophora* is a relatively small thin-shelled helmet shell. It weighs up to 80 g without shell (own estimation). Little is known about its feeding biology (Fig. 2).

Fig. 2. Two sulfuric acid-secreting marine molluscs. *Left Tonna galea* (Giant tun shell). *Right Cassidaria echinophora.* The shells were obtained at the Stazione Zoologica, Napoli, Italy. (The Tonna shell was a gift by Dr. Peter Dohrn). Photograph of empty shells. Scale = 10 cm

Cymatiidae (Tritons or Trumpet shells). *Charonia (Triton)* species are largesized predators of echinoderms. *Charonia tritonis* (Giant trumpet shell) feeds on the voracious Crown of Thorns *((Acanthaster planci)*, a starfish which is supposed to destroy coral reefs by feeding upon the polyps. *Cymatium nicobaricum* and related species are considerably smaller than *Charonia.* They attack and feed upon other gastropods (Houbrick and Fretter 1969). *Argobuccinum argus,* a South African whelk, feeds upon the polychaete *Gunnera capensis* (Day 1969). The feeding habits of another species, *Fusitriton (Argobuccinum) oregonensis* do not seem to be known.

Ficidae (Fig shells). *Ficus* species have a thin shell. They are active animals which live in sand and feed on sea urchins or other echinoderms (Wilson and Gillett 1974). They live in tropical or warm, temperate seas.

Bursidae (Frog shells). The frog shell family comprises gastropods closely related to the Cymatiidae. *Bursa granularis* feeds on polychaete and sipunculid worms (Houbrick and Fretter 1969).

3 The Sulfuric Acid Secretion of Tonnaceans

In 1853 the well known biologists F.H. Troschel and Johannes Müller were working together for a few weeks in a provisional laboratory in Messina investigating marine animals, mostly associated with the upwelling phenomenon of the Stretto di Messina. Professor Troschel had bought two live specimens of the Giant tun shell, *Tonna galea,* from fishermen in the harbour. In order to examine their anatomy he started to break one of the shells. The mollusc immediately stretched out its proboscis, "1 inch

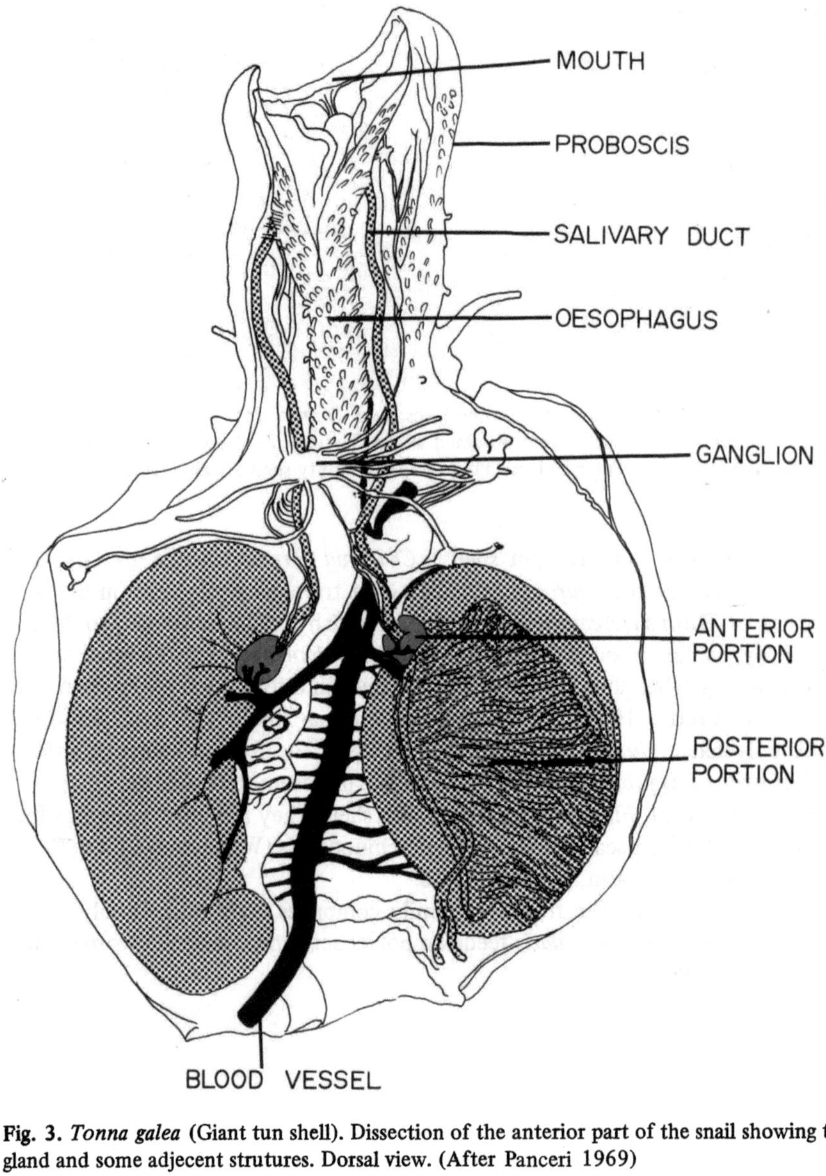

Fig. 3. *Tonna galea* (Giant tun shell). Dissection of the anterior part of the snail showing the acid gland and some adjecent strutures. Dorsal view. (After Panceri 1969)

thick and 6–7 inches long", and squirted from its tip a jet of clear fluid which fell on the marble floor. A lively effervescence appeared which demonstrated that the secretion was intensely acid. The excited professor shouted to his collegue, the physiologist Johannes Müller, to come and see the phenomenon. Samples of the secretion were collected, and analysed in Germany. They contained absout 3% sulfuric acid (Troschel 1854). Later analyses of the saliva of another tonnacean gastropod, *Cassidaria echinophora,* provided results in good agreement with the analyses made more than 120 years earlier on samples from *Tonna galea* (Fänge and Lidman 1976).

Most tonnacean gastropods probably have the ability to secrete a fluid from their salivary (buccal, proboscis) glands containing free H_2SO_4 (Panceri 1869; Day 1969; Houbrick and Fretter 1969; Fänge and Lidman 1976).

The structure of the salivary glands of *Tonna galea* and a few other sulfuric acid-secreting gastropods has been examined by Panceri (1869), later by Weber (1927), and Nüske (1973) studied the histology and ultrastructure of the salivary glands of *Cassidaria echinophora.* The glands are situated in the haemocoel on both sides of the oesophagus. From each gland a duct goes forwards, opening into the digestive tube close to the mouth. The gland of each side consists of two parts, one remarkably large posterior one, and one smaller anterior portion. The same general structure was found by Panceri (1869) in *Tonna galea* and other species (Fig. 3). The posterior portion of the *Cassidaria* salivary gland consists of transparent tubules closely packed in parallel to each other. The acid produced within the tubules is stored within distal vacuoles of large tubular cells. In their basal region the cells contain nuclei, mitochondria and Golgi membranes. The anterior glandular portion consists of two kinds of cells, mucus cells and canaliculi cells. According to Nüske (1973) the latter cells structurally resemble the parietal cells of mammalian gastric glands. The acid-producing vacuolated tubular cells show a strong resemblance with acid -producing epidermal and subepidermal gland cells of certain nudibranchs (Thompson 1983).

The biochemical mechanism of sulfuric acid production is not understood. Probably the secretion of sulfate ions is associated with the rich occurrence in marine molluscs of heparin-like polysaccharides containing a large percentage of sulfate groups. Sulfated mucopolysaccharides have been described from the hypobranchial glands of the trumpet shell *Charonia lampas* (Soda and Egami 1938) and the Atlantic common whelk, *Buccinum undatum* (Hunt and Jevons 1965), and in the eggs of *Tonna galea* (Horstmann 1960). Incorporation of sulfate into charonin-sulfuric acid, a polysaccharide from *Charonia lampas,* requires presence and interaction of "active sulfate" (3'-phosphoadenosin-5'-phosphosulfate). This is formed from ATP and sulfate under the influence of enzymes contained in the hypobranchial gland of *Charonia lampas* (Yoshida 1963). Sulfuric acid is released from sulfated polysaccharides, when these are hydrolysed by sulfatases (Lloyd and Stuart 1966). Highly sulfated mucopolysaccharides are thought to be responsible for the low pH of the skin mucus of an echinoderm, *Ophiocoma nigra* (Fontaine 1964). However, the processes responsible for the concentration of protons within the secretions of extremely low pH in tonnacean salivary glands have still to be explained.

4 Evidence of Toxins in Acid Secretions

There are indications that acid secretions from marine gastropods cause toxic effects which cannot be explained solely by the low pH. Corman (1963) reported that initially neutralized saliva from *Cassis tuberosa* is toxic to the sea urchin *Diadema*. The active principle seems to be a neurotoxin which inactivates sensory receptors or afferent nerves. Saliva of *Charonia tritonis* inactivates thestarfish*Acanthaster planci*, and that of *Charonia rubicunda* paralyses the starfish, *Patriella regularis* (Endean 1972). Houbrick and Fretter (1969) reported that the acid secretions of *Cymatium* and *Bursa* species inactivate gastropods and polychaetes upon which these gastropods prey. Day (1969) found that the saliva of *Argobuccinum argus*, which has a pH of 1.1, contains a neurotoxin that abolishes sensory responses of various invertebrates. Dilutions of the saliva in the order of 1 in 5000 buffered to pH 7,4 narcotized sea urchins, gastropods and polychaete worms.

Schoenlein (1898) and Henze (1901) found that the acid secretion of the salivary glands of *Tritonium (Charonia) nodiferum* contains an organic acid, aspartic acid, instead of sulfuric acid, but this has not been verified by any later investigations. Asano and Itoh (1960) found that the salivary glands of *Fusitriton (Argobuccinum) oregonensis* produce tetramine or tetramine-like compounds. Tetramine is the main component of the toxic secretion of the salivary glands of *Neptunea* species which

Table 1. Relationships between groups of gastropods mentioned in the article. (Systematics after Thiele, 1963)

Order	Family or superfamily	Genus
Prosobranchia		
Arachaegastropoda	Turbinidae	*Turbo*
Mesogastropoda	Naticidae	*Natica*
	Cypraeidae	*Cypraea*
	Tonnacea	*Argobuccinum, Bursa, Cassidaria, Cassis, Charonia, Cymatium, Ficus, Tonna*
	Muricidae	*Murex, Nucella, Thais, Urosalpinx*
	Buccinidae	*Babylonia, Buccinum, Neptunea*
	Conidae	*Conus*
Opisthobranchia		
Tectibranchia	Philinidae	*Philine*
	Aplysiidae	*Aplysia*
Notaspidea	Pleurobranchidae	*Pleurobranchus*
Nudibranchia	Dorididae	*Doris*

belong to the family Buccinidae, another group of marine gastropods (Table 1). The secretions from the neptuneid whelks are toxic, but not acid.

Narita et al. (1981) have reported food poisoning due to ingestion of a trumpet shell, *Charonia sauliae*. They suggested that the poisoning might have been due to tetrodotoxin contained in the digestive system. It was not clear whether the toxin was produced by the snail or came from some exogenous source.

Muricidae (Purple whelks). This is a family of marine carnivorous snails with numerous species. The shells are often covered with spines. Many muricids have the habit of attacking other molluscs or sessile crustaceans (barnacles) by drilling holes into their shells. The muricids produce a slightly toxic slime by their hypobranchial gland, and use another gland called the accessory boring organ as an aid in boring.

Hypobranchial (purple) gland. Tyrian purple was a dye made from the nasty smelling fluid collected from dying muricid whelks, expecially *Murex trunculus*. The ancient phoenicians were experts in making this dye. Archeologist have been able to trace the migrations of phoenicians around the Mediterranean by looking for heaps of empty *Murex* shells indicating sites of ancient purple-dye factories. Sometimes the phoenicians overfished their *Murex* banks and had to move to new locations in order to fish the whelks (Glover 1957; McCord 1969). The use of Tyrian purple lasted for centuries, but in 1907 Friedländer isolated, crystallized and analysed purple from 12,000 specimens of *Murex brandaris* and discovered that the dye was identical with the compound dibromindigo, which could be synthesized.

In the living purple whelks the purple exists in the form of colourless precursors such as 6-bromo-2-methylthioindoxyl (Baker and Sutherland 1968). The precursors are transformed into purple by sulfatases when exposed to ultraviolet light (sunligth).

In addition to purple precursors the hypobranchial gland of *Murex* species contains a series of biologically active compounds such as murexine or urocanylcholine [beta-imidazole-4(5)-acryl-choline] and 5-hydroxytryptamine (5-HT) Erspamer and Glässer 1957). Urocanylcholine has neuromuscular blocking activity and shows nicotinic but not muscarinic cholinergic effects on living tissues and cells. Urocanylcholine has been found in the hypobranchial gland of *Murex trunculus*, and also in *Urosalpinx cinereus* and *Thais lapillus*, but *T. floridana* and *T. chocolata* contain senecioylcholine instead (Whittaker 1960; Roseghini et al. 1970), and *T. haemastoma* contains dihydromurexine (imidazolepropionylcholine). The hypobranchial gland of *Concholepas concholepas* was found to contain urocanylcholine, choline, urocanic acid and ethyl urocanate (Roseghini et al. 1970). In *Nucella emarginata* and *Acanthina spirata* N-methyl murexine was tentatively identified (Bender et al. 1974).

The biological importance of the various biologically active compounds produced by the muricid hypobranchial gland remains a mystery. However, Hemingway (1978), who found a paralyzing substance – possibly a choline ester – in the hypobranchial gland of *Acanthina spirata*, suggested that the substance may assist the whelk in overwhelming its prey by interfering with neuromuscular functions in the latter.

Accessory boring organs (ABO). Muricids, and naticids (Table 1), utilize their radula apparatus as a drill by which they are able to drill holes through the calcareous shells of other invertebrates, mainly molluscs. A small dishshaped gland, named the accessory boring organ (ABO), is situated at the tip of the proboscis close to the mouth.

The ABO is supposed to produce a secretion which softens the shell of the prey. Carriker et al. (1967) by using microelectrodes succeeded in measuring the pH at the site of boring and found a minimum pH of 4.8. The cells of ABO contain cytochrome oxidase, succinate dehydrogenase and lactate dehydrogenase indicating an active aerobic metabolism (Person et al 1967). Chétail and Fournié (1969) showed that the ABO of *Thais lapillus* contains high activity of the enzyme carbonic anhydrase, which seems to be necessary for the production of H^+ ions during boring. When carbonic anhydrase in inhibited by Diamox, boring is inhibited, while administration of CO_2, or a mixture of CO_2 and O_2, accelerates boring.

A shell-boring behaviour, similar to that of muricids, was studied in *Natica* by Ankel (1937).

Buccinidae (Whelks). The family contains at least 2000 species (Wilson and Gillet 1974). The buccinids are carnivorous, often scavengers but they also take living preys.

Neptunea species. *Neptunea antiqua* (Red whelk) occurs in the same North-atlantic waters as the smaller-sized species *Buccinum undatum* (the Common whelk). It weighs up to 80 g minus the shell (own estimations). The Red whelk feeds upon living polychaetes but may accept dead fish or mussels when kept in captivity (Pearce and Thorson 1967). According to Shimek (1983, pers. comm.) neptuneid gastropods from Alaska are primarily predators on live polychaetes and bivalves.

In contrast to *B. undatum*, which has small salivary glands, *N. antiqua* has large salivary glands weighing up to 3–5% of the body weight (Fänge 1960). The glands of *Neptunea* species contain large amount of tetramine (tetramethylammonium salts) and low concentrations of choline, choline esters and histamine (Asano and Itho 1960; Fänge 1957, 1960). Tetramine has strong cholinergic effects on vertrebrate organs and shows a curare-like action (Ackermann 1921). Extracts of the glands are highly toxic to fish, mice and other animals. Ingestion of *N arthritica* (Asano 1952) and *N. antiqua* (Fleming 1971; v. Hofsten 1983, pers. commun.) have caused acute poisoning in man. Fänge (1960) suggested that *N. antiqua* uses its tetramine glands for defense and for paralyzing its preys. However, Pearce and Thorson (1967) in observations on living specimens found no evidence of any use of the venom in feeding activities.

Except in *Neptunea* species *(N. arthritica, N. intersculpta)*, according to Asano and Itoh (1960), poisonous salivary glands probably contraining tetramine are found in *Fusitriton oregonensis* and *Buccinum leucostoma.*

Histochemistry of Neptunea tetramine-glands. The *Neptunea* salivary glands are difficult to study with ordinary histological methods, because the tetramine-producing cell structures are destroyed by ordinary fixation fluids. The tetramine does not occur in free state in the cells but is stored within minute precursor granules containing acid polysaccharides. When the gland cells are activated by various stimuli, tetramine is instantly produced, probably by hydrolysis and transmethylation of some cellular components. The release of tetramine takes place very quickly. Because this amine is extremely soluble in water containing inorganic salts (as cell and tissue fluids), the appearance of tetramine is accompained by almost explosive uptake of water. This osmotic swelling breaks down the normal cell structures (Fänge 1960). Berg Justesen (1966) found that the tetramine precursor granules may be preserved in air -dried frozen sections fixed in 100% ethanol and slightly stained with thionine blue in 90% ethanol, or by a procedure using the quaternary ammonium compound, cetyltrimethylammonium bromide (C.T.A.B.) to prevent the release of tetramine.

Tetramine is an important component of coelenterate toxins (Welsh and Prock 1958) Berg Justessen (pers. commun) has suggested that tetramine might have something to do with the "explosion" of coelenterate nematocysts (nettle capsules), when these are touched or chemically stimulated. However, tetramine may not occur in all coelenterates provided with nematocysts.

Tetramine is chemically related to choline compounds, such as those occuring in the hypobranchial glands of muricids. A sulfur contraining ammonium derivative, 3-methylthiopropyltrimethylammonium chloride, isolated from the digestive system of the gastropod *Turbo argystomum* (fam. Turbinidae), probably is derived from the diet rather than being synthesized by the snail (Yasumoto and Endo 1974). *Babylonia japonica* (Japanese ivory shell; fam Buccinidae) has given rise to poisoning when ingested. An antinicotinic and ganglia blocking substance, "surugatoxin", was isolated from this species. In anaesthetized cat the toxin caused a prolonged fall of blood pressure. Tetrodotoxin and a metabolite of this were also found in *B. japonica* (Hayashi and Yamada 1975; Yasumoto et al. 1981).

5 Conacea (Toxoglossa) (Cone Shells and Related Forms)

The fam. Conidae comprises one genus, *Conus*, with a few hundred species living in tropical and warm temperate seas. Many cones have very decorative shells. The extremely rare "Glory of the Sea" *(Conus gloriamaris)* is famous among shell collectors. The cones and gastropods of related families (Turridae, Terebridae) feed on living preys, which are paralyzed by injected venom. The venom is transferred into the prey by a modified radula tooth, which consists of a thin chitinous sheet rolled into a little hollow structure strikely similar to a harpoon or dart. Different *Conus* species feed on worms (vermivorous), molluscs (molluscivorous) or fishes (piscivorous) (Kohn et al 1960). Most common are the vermivorous cones feeding on polychaetes. In accordance with the type of feeding, *Conus* species differ as regards to the structure of the radula teeth and the chemical composition of the venom. Venoms from molluscivorous cones have no effect on vertebrates and venoms from vermivorous species have only weak effects. On the other hand venoms from piscivorous cones are very toxic to fishes and other vertebrates (Endean 1972). Bites from *Conus geographus*, and perhaps one or two other piscivorous species, have caused death in man.

The *Conus* venoms contain a complex of substances: amines, quaternary ammonium compounds, peptides, proteins (Kohn et al. 1960; Halstead and Halstead 1978; Kobayashi et al. 1982 a). Vasoactive and cardioactive proteins occur in venoms of *C. eburneus* and *C. striatus* (Kobayashi et al. 1982 a, b). The amino acid sequences and the pharmacological properties of three toxic peptides isolated from *C. geographus* have been investigated (McManus et al. 1981; Gray et al. 1981). These peptides block neuromuscular transmission in vertebrate nerve-muscle preparations by interfering with acetylcholine receptors. The venom of *C. magus* (a piscivorous species) contains a 14-residue peptide with an amino acid sequence homologous with those of peptides from *C. geographus* (McIntosch et al. 1982). A 22-residue peptide isolated from the venom of *C. geographus* has an unusual amino acid composition (Stone and Gray 1982).

A radular apparatus resembling that of *Conus* species is found in gastropods of the families Terebridae and Turridae. These gastropods, reported to feed on live polychaetes etc., probably are provided with venom producing glands similar to those of cones (Shimek and Kohn 1981).

6 Opisthobranchia (Sea-Slugs, Sea Hares)

The opisthobranchs comprise about 3000 species most of which lack the shell or have a highly reduced one. However, many opisthobranchs have evolved defensive mechanisms which may compensàte for the lack of a protecting shell. In *The Log from the Sea of Cortez* the author John Steinbeck tells that his friend, the marine biologist Ed Rickets, wondering why nudibranchs are never eaten by fishes, put a living nudibranch into his mouth. He then made a wry face, spat and vomited but had the satisfaction of knowing why. Nudibranchs, and several other opisthobranchs, make use of offensive skin secretions to protect themselves against predators. A sophisticated method is used by certain species of opisthobranchs, which do not secrete venoms produced by themselves, but use coelenterate nematocysts from their diet, as weapons. For example, the planktonic *Glaucus atlanticus* utilizes the highly venomous nettle capsules from *Physalia* (Portuguese Man o'War), which it stores in skin protuberances (Thompson 1976). Other species, as observed already by Panceri (1869), produce an acid secretion from epidermal and subepidermal glands *(Pleurobranchus, Berthella, Philine)*. The acid secretory cells are cylindrical with large distal vacuoles containing sulfuric acid (Edmunds 1968; Thompson 1976, 1983). The vacuolar contents are released when the animals are abruptly disturbed. Similar skin glands are also found in certain marine prosobranch gastropods (Cypraeaceae, Lamellariaceae) (Kniften 1968).

In addition to acids opisthobranch skin glands produce various toxic organic compoinds. Johannes (1963) reported that the nudibranch *Phyllidia varicosa* secretes a toxic mucus active on crustaceans and fish. Fenical et al. (1979) demonstrated in the cephalaspid optisthobranch *Navanax inermis* a bright yellow secretion from a gland near the anus. The secretion contains low-molecular compounds termed navanones. The biological effects from these compounds are little known.

Sea hares (*Aplysia* species) secrete offensive fluids from a purple gland and from a multiporous opaline gland. The "purple" produced by *Aplysia* species consists of chromoproteins. Heterocyclic sesquiterpene bromo-compounds, "aplysin" and "aplysinol", isolated from *Aplysia,* show distant relationships with the precursors of Tyrian purple of muricid snails (Yamamura and Hirata 1963). Toxic factors from *Aplysia* digestive glands have been characterized by Winkler et al. (1962) and Watson and Rayner (1973). The sea hare toxins are assumed to be largely of dietary origin. Thus a dibromoaplysiatoxin from the Hawaian sea hare, *Stylocheilus longicauda,* has also been isolated from blue-green algae. Certain algal metabolites found in sea hares may have antileukemia activity (Faulkner 1977; Scheuer 1977).

Fuhrman et al. (1979) investigated toxins from digestive glands of dorid nudibranchs. An active fraction designated "dorid toxin" produced vascular responses in rat and negative inotropic and chronotropic effects on isolated hearts. The toxin had a molecular weight under 8000 and was heat-stable. Because it was not destroyed by proteases, it is doubful whether it is a peptide.

7 Cephalopods (Octopods, Squids, Cuttlefish)

Cephalopods probably are the highest organized invertebrate animals. In several respects they have developed in parallel to vertebrates. They are free swimming predators possessing a head and a complex brain, large eyes and a closed blood (hemolymph) circulation showing vasomotor activities (Bacq and Ghiretti 1953).

In 1888 Salvatore Lo Bianco at the Stazione Zoologica in Napoli, Italy, discovered that the posterior salivary gland of *Octopus* contains a substance toxic for crabs. The posterior salivary gland secretion is strongly acid *(Octopus, Sepia)*. A long series of physiologically active low-molecular compounds have been isolated from cephalopod venom glands: tyramine, octopamine (p-hydroxyphenylethanolamine), dopamine, noradrenaline, 5-HT, histamine, taurine, acetylcholine (Hartman et al. 1960). The components vary very much from species to species. The glands also contain proteolytic enzymes and hyaluronidase (Ghiretti 1960). Erspamer and coworkers found a physiologically active peptide, named eledoisin, in the posterior salivary gland of *Eledone moschata.* It is hypotensive, stimulates smooth muscles and causes an atropine-resistant increase in salivary secretion in mammals (Erspamer 1981). The biological activity of eledoisin is destroyed by chymotrypsin, or tissue extracts, indicating its peptidic nature.

However, there is no evidence that either amines, or the peptide eledoisin, occur in quantities enough to explain the crustacean-killing activity of octopod venoms. According to Ghiretti (1960) the killing effect of extracts of the posterior salivary glands of *Octopus vulgaris* is due to a glycoprotein, "cephalotoxin". Cariello and Zanetti (1977) fractionated cephalotoxin into five components, two of which were acidic glycoproteins, which were named alpha- and beta-cephalotoxin respectively. Their molecular weights were 91,200 and 33,900. Songdahl and Shapiro (1974) purified a toxin with the molecular weight of about 23,000 from the posterior salivary gland of *Octopus dofleini.* Kem and Scott (1980) found a hemolytic protein in squid *(Loligo peali)* posterior salivary glands. Hoskin and Prusch (1983) characterized an enzyme, a DFP-ase, in the same organ.

The australian Blue-ringed octopus, *Octopus (Hapalochlaena) maculosus,* has remarkably large posterior salivary glands containing a venom dangerous to man. The active principle has a molecular weight lower than 540 (Sutherland et al. 1970). The physiological actions resemble those of tetrodotoxin (Gage et al. 1976; Freeman 1976). Sheumack et al. (1978) concluded that the toxin may be identical to tetrodotoxin.

The octopod, *Tremoctopus violaceus*, in similarity to certain opisthobranch gastropods, uses coelenterate toxins, from *Physalia* tentacles, as weapons (Jones 1963).

8 Discussion and Conclusions

Molluscan venoms have biological functions in association with feeding and in defense. Many gastropods, belonging to families not mentioned previously in this article, are carnivorous predators and scavengers: Fasciolariidae, Galeodidae, Magilidae, Olividae, Vasidae, Volutidae. Some of those forms may also have developed chemicals used for overwhelming live preys. However, the feeding biology of most marine molluscs is poorly known.

Investigations of toxins contained in molluscan tissues raise problems concerning the modes of biosynthesis. In most cases the toxins may be produced by cells of the molluscs themselves, but not infrequently powerful toxins might have been taken up from the environment, primarily from the diet. The tetramine glands of certain marine whelks (Buccinidae), and the hypobranchial glands of buccinid and muricid whelks, may be organs of choice for research on biosynthesis of cholinergic compounds. The mechanisms responsible for the production of exetremely low pH in salivary and skin secretions of certain marine gastropods are interesting but poorly understood.

Venoms occur only in a limited number of molluscan spcecies, but all molluscs contains biologically active substances which have roles in the normal physiology of these animals. Thus the sulfur-containing compound taurine, present in large amounts in marine molluscs, is important in intracellular osmoregulation (Lange 1963; Pettit et al. 1973). The chemically related isethionic acid occurs in cephalopod neurons (Clark et al. 1960). Prostaglandin peroxidase regulates reproduction in the abalone *(Haliotis rufescens)* (Morse et al. 1977). Amines and peptides found in molluscan nervous systems may be neurotransmittors. Antiviral and antibacterial substances have been detected in marine molluscs and so on (Pettit et al. 1970; Ruggieri 1976).

Certain marine molluscs, especially large-sized forms, are rare and should be protected. Work on substances contained in marine molluscs may inspire further fruitful research, but the availability of many species is too limited to allow the animals to be used as raw materials for useful drugs, unless aquaculture methods are applied.

Acknowledgements. I am grateful to Dr. Anders Warén (University of Göteborg, Sweden) and Dr. Ronald Shimek (Friday Harbor Lab., USA) for discussions on gastropod systematics and biology.

References

Ackermann D (1921) Kurze Bemerkungen über Curare-Ersatzpräparate. Münch Med Wochenschr 1: 12

Ankel WE (1937) Wie bohrt Natica. Biol Zentralbl 57: 75–82

Asano M (1952) Studies on toxic substances contained in marine animals. I. Locality of the poison of *Neptunea (Barbitonia) arthritica* (Bernardi). Bull Jpn Soc Sci Fish 17: 283–287

Asano M, Itoh M (1960) Salivary poison of a marine gastropod, *Neptunea arthritica* (Bernhardi) and the seasonal varistion of its toxicity. Ann NY Acad Sci 90: 674–688

Bacq ZM, Ghiretti F (1953) physiologie des glandes salivaries postérieures des Céphalopodes Octopodes isolées et perfusées in vitro. Pubbl Staz Zool Napoli 23: 267–277

Baker JT Sutherland MD (1968) Pigment of marine animals. VIII. Precursors of 6,6'-dibromindigotin (Tyrian purple) from the mollusc *Dicathais orbita* (Gmelin). Thetrahedron Lett (1): 43–46

Baslow MH (1977) Marine Pharmacology. Reprint (supplementary material). Krieger, Huntington, NY, p 360

Bender JA, DeRiemer K, Roberts TE, Rushton R, Boothe P, Mosher HS and Fuhrman FA (1974) Choline esters in the marine gastropods *Nucella emarginata* and *Acanthina spirata*: a new choline ester, tentatively identified as N-methyl-murexine. Comp Gen. Pharmacol 5:191–198

Berg Justesen N-P (1966) The demonstration of tetramine producin cells in the salivary glands of the marine gastropod *Neptunea antiqua*. Nytt Mag Zool (Oslo) 13: 33–38

Cariello L, Zanetti L (1977) α- and β-cephalotoxin: two paralyzing proteins from posterior salivary glands of *Octopus vulgaris*. Comp Biochem Physiol 57C: 169–173

Carriker MR, Charlton V, Van Zandt D (1967) Gastropod *Urosalpinx*: pH of accessory boring organ while boring. Science (Wash DC) 158: 920–922

Chêtail M, Fournié J (1969) Shell-boring mechanism of the gastropod *Purpura (Thais) lapillus*: a physiological demonstration of the role of carbonic anhydrase in the dissolution of Ca C03. Am Zool 9: 983–990

Clark WG, Hartman WJ, Liebhold RA, Jordon AL, Cyr SD (1960) Some aspects of the biochemical pharmacology of the Octopus. Proc West Pharmacol Soc 3: 106–122

Corman I (1963) Toxic properties of the saliva of *Cassis*. Nature (Lond) 200: 88–89

Day JA (1969) Feeding of the Cymatiid gastropod *Argobuccinum argus*, in relation to the structure of the proboscis and secretions of the proboscis gland. Am Zool 9: 909–916

Edmunds M (1968) Acid secretion in some species of Doridacea (Mollusca, Nudobranchia). Proc Malacol Soc Lond 38: 121–133

Endean R (1972) Aspects of molluscan pharmacology. In: Chemical Zoology, Vol. 7. Academic, New York, pp 421–466

Erspamer V (1981) The tachykinin peptide family. TINS 4: 267–269

Erspamer V, Glässer A (1957) The pharmacological actions of murexine (urocanylcholine) Br J Pharmacol 12:176–184

Fänge R (1957) An acetylcholine-like salivary poison in the marine gastropod *Neptunea antiqua*. Nature (Lond) 180: 196–197

Fänge R (1960) The salivary gland of *Neptunea antiqua*. Ann NY Acad Sci 90: 689–694

Fänge R, Lidman U (1976) Secretion of sulfuric acid in *Cassidaria* echinophora (Lamarck) Mollusca: Mesogastropoda, marine carnivorous snail). Comp Biochem Physiol 53 A: 101–103

Faulkner DJ (1977) Interesting aspects of marine natural products chemistry. Tetrahedron 33: 1421–1443

Fenical W, Sleeper HL, Paul VJ, Stallard MO, Sun HH (1979) Defensive chemistry of *Navanax* and related opistrobranch molluscs. Pure Appl. Chem 51: 1865–1874

Fleming C (1971) Case of poisoning from Red whelk. Br Med J 3: 520:521

Fontaine AR (1964) The integumentary mocous secretions of the ophiuroid *Ophiocoma nigra*. J Mar Biol Assoc UK 44: 145–162

Freeman SE (1976) Electrophysiological properties of maculotoxin. Toxin 14: 396–399

Friedländer P (1907) Recognition of the dye of antique purple from *Murex brandaris*. Mh Chem 28: 991–996

Fuhrman FA, Fuhrman GJ, DeRoemer K (1979) Toxicity and pharmacology of extracts from dorid nudibranchs. Biol Bull (Woods Nole) 156: 289–299

Gage PW, Moore JW, Westerfield M (1976) An octopus toxin, maculotoxin, selectively blocks sodium current in squid axons. J Physiol (Lond) 259: 427–443

Ghiretti F (1960) Toxicity of *Octopus saliva* against crustacea. Ann NY Acad Sci 90: 726–741

Glover TR (1957) The ancient world. Penguin Books Harmondsworth, Middlesex, UK, p 240

Gray WR, Luque A, Olivera BM, Barrett J, Cruz LJ (1981) Peptide toxins from *Conus geographus* venom. J Biol Chem 256: 4734–4740

Halstead BW, Halstead LG (1978) Poisonous and venomous marine animals of the world (revised edition). Darwin, Princeton, New Jersey, p 1326

Hartman WJ, Clark WG, Cÿr SD, Jordon AL, Leibhold RA (1960) Pharmacologically active amines and their biogenesis in the octopus. Ann Acad Sci 90: 637–666

Hayashi E, Yamada S (1975) Pharmacological studies on surugatoxin, the toxic principle from Japanese ivory mollusc *(Babylonia japonica)*. Br J Pharmacol 53: 207–215

Hemingway GT (1978) Evidence for a paralytic venom in the intertidal snail *Acanthina spirata* (Neogastropoda: Thaisidae). Comp Biochem Physiol 60C: 79–81

Henze M (1901) Über das vorkommen freier Asparaginsäure im tierischen Organismus. Ber d chem Ges Berl 34: 348–354

Horstmann HJ (1960) Untersuchungen über Polysaccharide aus den Eiern einiger mariner Mollusken. Pubbl Zool Napoli 31: 308–319

Hoskin FCG and Prusch RD (1983) Characterization of a DFP-hydrolyzing enzyme in squid posterior salivary gland by use of Soman, DFP and manganous ion. Comp Biochem Physiol 75C: 17–20

Houbrick JR, Fretter V (1969) Some aspects of the functional anatomy and biology of *Cymatium* and *Bursa*. Proc Malac Soc Lond 38: 415–429

Hughes RN, Hughes HPI (1971) A study of the gastropod *Cassis tuberosa* (L.) preying upon sea urchins. J Exp Mar Biol Ecol 7: 305–314

Hunt S, Jevons FR (1965) The polysaccharide sulphate-peptide complex in the hypobranchial mucin of *Buccinum undatum* (L.) Biochim Biophys Acta 101: 214–216

Johannes RE (1963) A poison-secreting nudibranch (Mollusca, Opisthobranchia). Veliger 5: 104–105

Jones EC (1963) *Tremoctopus violaceus* uses *Physalia* tentacles as weapons. Science (Wash DC) 139: 764–766

Kniffen JC (1968) Acid secretion in *Cypraea*. Am J Dig Dis 13: 775–778

Kem WR, Scott JD (1980) Partial purification and characterization of a cytotoxic protein from squid *(Loligo pealei)* posterior salivary glands. Biol Bull (Woods Hole) 159: 475

Kobayashi J, Nakamura H, Hirata Y, Ohizumi Y (1982a) Isolation of a cardiotonic glycoprotein stria toxin from the venom of the marine snail *Conus striatus*. Biochem Biophys Res Commun 105: 1389–1395

Kobayashi J, Ohizumi Y, Nakamura H, Hirata Y (1982b) Pharmacological study on the venom of the marine snail *Conus textile*. Toxicon 19

Kohn AJ, Saunders PR, Wiener S (1960) Preliminary studies on the venom of the marine snail *Conus*. Ann NY Acad Sci 90: 706–725

Lange R (1963) The osmotic function of amino acids and taurine in the mussel *Mytilus edulis*. Comp Biochem Physiol 10: 173–179

Lloyd PF, Stuart CH (1966) Acidimetric assay of glycosulphatase from *Patella vulgata*, acting on some monosaccharide sulphate, using a pH-stat. Biochem J 99: 37

McCord CP (1969) The lowly whelk and the lofty royal purple dye. Arch Environ Health 18: 379–385

McIntosh M, Cruz LJ, Hunkapiller MW, Gray WR, Olivera BM (1982) Isolation and structure of a peptide toxin from the marine snail *Conus magus*. Arch Biochem Biophys 218: 329–334

McManus OB, Musick JR, Gonzales C (1981) Peptides isolated from the venom of *Conus geographus* block neuromuscular transmission. Neurosci Lett 25: 57–62

Morse DE, Duncan H, Hooker N, Morse A (1977) Hydrogen peroxide induces spawning in molluscs with activation of prostaglandin endoperoxide synthetase. Science (Wash DC) 196: 298–300

Narita H, Noguchi T, Maruyama J, Ueda Y, Hashimoto K, Watanabe Y, Hida K (1981) Occurrence of tetrodotoxin in a trumpet shell boshubora *Charonia sauliae*. Bull Jpn Soc Sci Fish 47: 935–942

Nüske H (1973) Cytology of the acid-secreting gland of the marine snail *Cassidaria echinophora*. Cytobiologie 7: 164–180

Panceri P (1969) Gli organi e la secrezione dell' acido solforico nei Gasteropodi, con un'appendice relativa ad altri organi glandolari dei medesimi. Atti dell' Academia delle Science Fisiche e Matematiche Napoli 4 (10): 1–56, 4

Pearce JB, Thorson G (1967) The feeding and reproductive biology of the red whelk *Neptunea antiqua* (L.). (Gastropoda, Prosobranchia). Ophelia 4: 277–314

Person P, Smarsh A, Lipson J, Carriker MR (1967) Enzymes of the accessory boring organ of the muricid gastropod *Urosalpinx cinerea follyensis.* I. Aerobic and related oxidative systems. Biol Bull (Woods Hole) 133: 401–410

Pettit GR, Day JF, Hartwell JL, Wood HB Jr (1970) Antineoplastic components of marine animals. Nature (Lond) 227: 962

Pettit GR, Ode RH, Harvey III TB (1973) Isolation of taurine from the molluscs *Macrocallista nimbosa* and *Turbo stenogyrus.* Lloydia (Cinci) 36: 204–206

Randall RE, Schroeder WA, Starck II (1964) Carib J Sci 4: 421 ff

Roseghini M, Erspamer V, Ramorino L, Gutierrez JE (1970) Choline esters, their precursors and metabolites in the hypobranchial gland of the prosobranchiate mollusc *Concholepas concholepas* and *Thais chocolata.* Eur Biochem 12: 468–473

Ruggieri GD (1976) Drugs from the sea. Science (Wash DC) 194: 491–497

Scheuer PJ (1973) Chemistry of marine natural products. Academic, New York

Scheuer PJ (1977) Marine toxins. Accounts of Chemical Research 10: 33–39

Schoenlein K (1898) Über eine Säuresecretion bei Schnecken. Biol 36: 523–548

Schroeder RE (1962) Urchin killer. Sea Frontiers 8: 156–160

Sheumack DD, Howden MEH, Spence I, Quinn RJ (1978) Maculotoxin: a neurotoxin from the venom glands of the octopus *Hapalochlaena maculosa* identified as tetrodotoxin. Sciene (Wash DC) 199: 188–189

Shimek RL, Kohn AJ (1981) Functional morphology and evolution of the toxoglossan radula. Malacologia 20: 423–438

Snyder N, Snyder H (1970) Alarm response of *Diadema antillarum*. Science (Wash DC) 168: 276–278

Soda T, Egami F (1938) On a new sulphuric acid ester in the mucus of *Charonia lampas.* Bull Chem Soc Jpn 13: 652

Songdahl JH, Shapiro BI (1974) Purification and composition of a toxin from the posterior salivary gland of *Octopus dofleini.* Toxicon 12: 109–115

Stone BL, Gray WR (1982) Ocurrence of hydroxy proline in a toxin from the marine snail *Conus geographus.* Arch Biochem Biophys 216: 765–767

Sutherland SK, Broad AJ, Lane WR (1970) Octopus neurotoxins: low molecular weight non-immunogenic toxins present in the salive of the blue-ringed octopus. Toxicon 8: 249–250

Thompson TE (1976) Biology of opisthobranch molluscs, Vol 1. The Ray Society, London, p 206

Thompson TE (1983) Detection of epithelial acid secretions in marine molluscs: review of techniques, and new analytical methods. Comp Biochem Physiol 74A: 615–621

Troschel FH (1854) Über den Speichel von *Dolium galea* (Lam.). Monatsber Königl Preuss Akad Wiss Berl 486–494

Watson M, Rayner MD (1973) Midgut gland toxins of Hawaian sea hares. II. A preliminary pharmacological study. Toxicon 11: 269–276

Weber H (1927) Der Darm von *Dolium galea* (L.) Eine vergleichende Untersuchung unter besonderer Berücksichtigung der *Tritonium*-arten. Z Morphol Ökol Tiere 8: 663–804

Welsh JH, Prock PB (1958) Quaternary ammonium bases in the coelenterates. Biol Bull (Woods Hole) 115: 551–561

Whittaker VP (1960) Pharmacologically actice choline esters in marine gastropods. Ann NY Aca Sci 90: 695–705

Wilson BR, Gillett K (1974) Australian shells. (revised edn). Reed, Syndey, AH, p 168

Winkler LR, Tilton BE, Hardinge MG (1962) A cholinergic agent extracted from sea hares. Arch Pharmacodyn (Ther) 137:76–83

Yamamura S, Hirata Y (1963) Structures of aplysin and aplysinol, naturally occurring bromocompounds. Tetrahedron 19: 1485–1496

Yasumoto T, Endo M (1974) Toxicity study on a marine snail, *Turbo argyrostoma* – II. Identification of (3-methylthiopropyl) trimethylammonium chloride. Bull Jpn Soc Sci Fish 40: 217–221
Yasumoto T, Oshima Y, Hosaka M, Miyakoshi S (1981) Occurrence of tetrodotoxin in the ivory shell *Babylonia japonica* from Wakasa Bay Japan. Bull Jpn Soc Sci Fish 47: 929–934
Yoshida H (1963) Formation de phosphoadénosine-phosphosulfate à partir de gland muqueuse de *Charonia lampas* et transfert de sulfate à l'acide charoninsulfate. Societé de Biologie 157: 1861–1863

Discharge Mechanism of the Nematocysts of *Pelagia noctiluca*

A. SALLEO[1]

1 Introduction

The nematocysts are complex organelles, contained in specialized cells, the nematocytes. Under adequate stimulus (Lubbock 1979; Tardent et al. 1980; Ertman and Davenport 1981) the nematocysts eject a threat that, in turn, either adheres to or penetrates into the prey, injecting the venomous substances contained into the capsule fluid. The nematocyst is generally believed to be formed as a secretory product of Golgi apparatus (Cormier and Hessinger 1980) although such an interpretation has been recently questioned (Holstein 1981). The complex process of differentiation which produces the three layers of the capsule wall and the thread has been recently described by Holstein (1981). The cnidocil apparatus seems to be the sensory receptor that elicits the discharge in the intact system in situ (Cormier and Hessinger 1980). Nevertheless, the discharging capacity is retained in isolated nematocysts, although it is less prompt than it is in situ. Since quantitative and detailed data concerning the effectiveness of stimuli cannot be obtained on nematocysts in situ, isolated capsules are frequently employed with the aim to investigate the mechanism of discharge, which, in spite of the numerous attempts to elucidate it, is still uncertain.

Three alternative theories (Mariscal 1974; Cormier and Hessinger 1980) have been proposed so far to explain the eversion of the thread: (a) the capsule fluid exerts a constant pressure on the capsule wall. When the operculum is loosened the thread is pushed out explosively. (b) The thread is everted by a sudden increase of inner pressure, which, in turn is caused by an influx of water. (c) The thread is everted as a consequence of activation of contractile structures. The latter theory appears to be less probable, since contractile structures have not been identified so far either in the nematocyte or in the nematocysts (Wood and Novak 1982). Theories (a) and (b) assume that the capsule wall is not permeable to water in the resting state. Recently, Lubbock and Amos (1981) have shown that the resting capsule of *Rhodactis rhodostoma* is permeable up to a molecular weight of 670. Salleo et al. (1983) observed that undischarged capsules of *Pelagia noctiluca* are permeable to methylene blue (M.W. 355,90). Therefore, since water and small solute can diffuse freely through the capsule wall, neither a constant pressure in the resting state, nor an abrupt increase in permeability to water should be involved in the discharge mechanism. On the other

---

1 Institute of General Physiology, University of Messina, Via dei Verdi 85, 98100 Messina, Italy

Toxins, Drugs, and Pollutants in Marine Animals
(ed. by Bolis et al.)

hand, an influx of water could be induced by an increase in number of osmotically active solutes in the capsule fluid. This mechanism has been proposed by Lubbock et al. (1981) and by Lubbock and Amos (1981). The physicochemical basis of this model is a very high concentration of proteins and of calcium in the capsule fluid. Calcium could aggregate protein molecules; when calcium is unbound, the protein is disaggregated so that the number of osmotically active solutes increases. Consequently water is drawn into the capsule fluid, raising the hydrostatic pressure to the discharge level. This model places the discharge mechanism into the capsule fluid. The chemical composition of the capsule fluid has been investigated by Blanquet (1968) on microbasic mastigophores of *Aiptasia pallida* and by Calabrese et al. (1983) on holotrichous isorhiza of *Pelagia noctiluca.* In both types a protein characterized by an unusually high content ($\simeq$ 80%) of glutamic acid has been found. The carboxyl groups of this protein could be stabilized by Ca^{2+}.

On the other hand, when the discharging action of proteolytic enzymes (Yanagita and Wada 1954; Salleo et al. 1983) is taken into account, the role of the capsule wall appears to be relevant. In fact, these enzymes cannot penetrate into the capsule fluid, owing to their molecular weight. Therefore they should act on the capsule wall. The chemical composition of the latter has been investigated by many authors on different types of nematocysts obtained from different species. The presence of keratin (Yanagita and Wada 1954), mucopolysaccarides (Hamon 1955), cartilagenous material (Phillips 1956) have been described. Successively, Blanquet and Lenhoff (1966) observed that the capsule of the nematocysts of *Aiptasia pallida* is mainly composed of a collagenous-like protein linked by disulfide bonds. This result has been successively confirmed on the basis of X-ray microanalysis ((Mariscal 1980). The dissolving action of dithiothreitol (DTE) has been tested by Mariscal and Lenhoff (1969) on a number of different nematocysts. These authors observed also that DTE was active only on discharged capsules, while it did not dissolve the resting ones. Moreover, some types of nematocysts were not dissolved at all. The fact that only discharged capsules were attacked by DTE could suggest, in my opinion, that some change in the capsule wall could be associated with discharge. Moreover, since some capsules were dissolved by DTE while others were not, the chemical composition of the capsule wall may differ in different nematocysts types.

Also the response to discharging agents such as neutral salts and pH is markedly different in different capsules. Blanquet (1970) observed that microbasic mastigophores of *Aiptasia pallida* are discharged in diluted sea water, and that they are extruded undischarged in concentrated citrate solutions. When the effect of neutral salts was tested on isolated nematocysts it was observed that the discharging effect increased with decreasing the ionic concentration. Therefore this type of nematocysts seems to be highly sensitive to osmotic gradient. The order of effectiveness of neutral salts was K_2SO_4, Na_2SO_4 > KCl, NaCl > KNO_3, $NaNO_3$.

On the other hand, the holotrichous isorhizas of *Pelagia noctiluca* are extruded undischarged in distilled water and retain their discharging capacity (Salleo et al. 1983). Therefore they are insensitive to osmotic gradient. To explain the lack of discharge in hypotonic solutions and in distilled water an impermeability of the resting capsule to water cannot be claimed, since these capsules were found to be freely permeable to methylene blue (Salleo et al. 1983). Morever, sulfate had little or no effect, while chloride was more effective as a discharging agent. I^- has been found to be very effective. In any case the discharging rate increased with increasing ionic concentrations (Salleo et al. 1983 a).

As regards the effect of pH, Yanagita and Wada (1953) described that the nematocysts of *Diadumene* discharged at pH < 3 and > 11. The discharging effect of bases was quite similar in the nematocysts of *Pelagia noctiluca* (Salleo et al. 1983 b) while on the other hand, the effect of low pH was noticeably different. In fact, at pH 3, instead of discharge, a collapse of the capsule wall was observed, which impaired permanently the discharging capacity. Only at pH 1 a small percentage, compared to that of wall-collapsed capsules, discharged. These results, apart from indicating the differences that can be found in different nematocysts, stress the role that the capsule wall may play in the process of discharge.

A common feature, which has been generally observed on different nematocysts is the inhibitory effect of Ca^{2+} and other divalent cations on the discharge process (Blanquet 1970; Lubbock and Amos 1981; Salleo et al. 1983 a). Such an effect has been interpreted to depend on stabilization of the conformation of a protein molecule.

From a general survey of the literature the following concepts can be stressed:

1. Different nematocysts show different responses to discharging agents such as osmotic gradient, anions and pH. This could be due to different chemical composition and biophysical characteristics of the capsule wall.
2. Divalent cations inhibit the discharge.
3. A conformational change in a protein molecule could play a relevant role in the process of discharge. Whether this protein is located into the capsule fluid or on the capsule wall is still uncertain.

The following experiments have been designed in order to elucidate the role of the capsule wall in the discharge process of holotrichous isorhiza nematocysts of *Pelagia noctiluca:* (a) a morphological study of the capsule wall at light and scanning electron microscope (S.E.M.) performed on resting, discharged, broken and collapsed nematocysts; (b) a comparison of permeability to dyes in normal and collapsed capsule wall; (c) a study on the inhibitory effect of polyvalent cations on the discharge induced by KI and trypsin.

2 Materials and Methods

The nematocysts were obtained from tentacles of *Pelagia noctiluca* 2 h after capture. The excised tentacles were submerged in distilled water for 2 h. The treatment with distilled water produces the detachment of the epidermis and the osmotic rupture of nematocytes, so that the nematocysts are delivered undischarged in water. The naked tentacle was removed, the suspension was repeatedly washed in distilled water and filtered through plancton nets in order to remove most tissue debris. Then the suspension was frozen at $-20^{o}C$. Before use the samples were defrosted and filtered again. The suspension, which contained the large holotrichous isorihzas, the smaller euryteles and very fine tissue debris were allowed to settle in a test tube. Since the holotrichous isorhizas settle more rapidly than the smaller nematocysts and debris, most of the latter can be removed from the sample.

2.1 Morphological Study

Three different samples were prepared. One sample contained normal undischarged nematocysts. In a population of normal undischarged capsules some broken capsules can be casually encountered. They can be easily recognized from the small slit from which the uneverted thread is hanging out. Since the thread has not been everted no spines can be observed on the outer surface of the thread. A second sample contained discharged nematocysts. Discharge was induced by treatment with Kl 0.533 M (Salleo et al. 1983 a) The third sample contained capsules whose wall had been induced to collapse by treatment with HCl. This effect of pH has been described elsewhere (Salleo et al. 1983 b). Briefly, when the nematocysts are treated with pH lower than 3 (Fig. 1) they lose their discharging capacity, owing to a collapse of the capsule wall. This phenomenon is also produced, at a slower rate, when the nematocysts are maintained in distilled water at room temperature for a few hours.

The three samples were fixed with buffered glutaraldehyde (pH 7) and dehydrated by treatment with a series of alchools. Then they were placed on specimen stubs, coated with gold-platinum and observed on Autoscan Siemens of E.T.E.C.

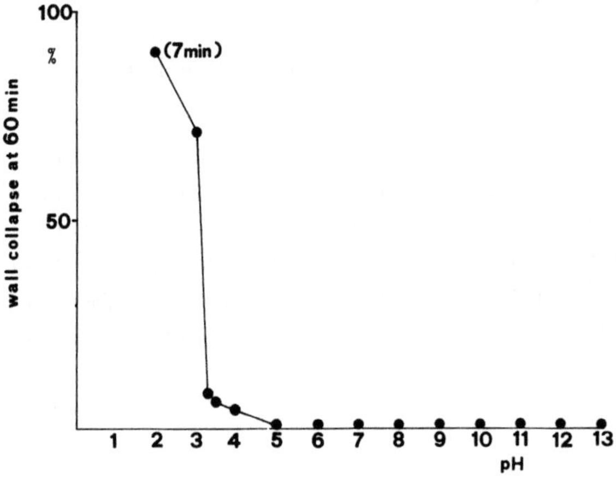

Fig. 1. Percentage of wall collapsed nematocysts by 60 min at different values of pH. Note that at pH 2 about 90% wall collapsed capsules were observed by 7 min

2.2 Permeability of Normal and pH Treated Capsules

In order to study which physiological changes are associated to the morphological ones in collapsed capsules, a test of permeability to dyes having different molecular weight has been performed. A control test has been made on normal undischarged capsules.

On the basis of the changes observed at S.E.M., on the pH treated capsules (see below) the Gram staining has been performed on normal undischarged, discharged and collapsed capsules. Moreover, the dissolving effect of lisozime (1%, pH 6.2) was also tested.

2.3 Effect of Polyvalent Cations

The inhibitory effect of a previous treatment with $ThCl_4$ on the discharge induced by I^- and trypsin has been studied. Th^{4+} is known to have relevant surface actions, so that, for instance it changes the electrophoretic mobility of proteins (Diamond and Wright 1969) and it reverses the streaming potential (Cignitti et al. 1970). Since Th^{4+}, although forming complex ions, can penetrate through the capsule wall, also the effect of Ruthenium red, a dye that cannot penetrate, owing to its molecular weight, and is known to bind to negative groups of cell wall, has been tested.

The nematocysts were suspended for a few minutes in 1, 5, 10, 20 mM $ThCl_4$ or in 1% Ruthenium red solutions. The solution was removed and the nematocysts were washed in distilled water. Then either 533 mM KI or 1% trypsin were applied. The discharged capsules were counted and the result was expressed as percentage of discharged capsules with respect to the total population of normal capsules initially present in the field. Control tests were performed by applying KI or trypsin to untreated nematocysts.

3 Results

3.1 Morphological Study

The volume of the normal undischarged capsules is about 40% larger than that of discharged ones (Fig. 2). Their shape differs slightly from spherical, being the largest the diameter which passes through the operculum (Fig. 7). This shape is not altered by dehydration and vacuum treatment that are common steps for a S.E.M. examination. Is is worthwhile to stress that also a dehydration performed on a glass slide by exposing the capsule to air does not produce a flattening of these capsules.

At S.E.M. observation the most relevant feature is a markedly rough surface (Fig. 3) as previously described (Salleo et al. 1983).

The discharged capsule, besides being smaller than the resting one, shows different response to dehydration with respect to the latter. In fact, light microscope observations reveal that when suspended in water they maintain their approximately spherical shape (Fig. 2), while under dehydration they tend to flatten. At S.E.M. observation, following dehydration and vacuum treatment the discharged capsules look like a deflated rubber balloon (Fig. 4). Such a difference respect to the resting capsules could be ascribed to the lack in the discharged ones of the content, i.e. the coiled thread and the capsule fluid. The observed shape of broken capsules contradicts this interpretation (Figs. 5, 6). The broken capsule is devoid of content, since the fluid has flowed out and the thread is hanging out from the lesion of the capsule. Nevertheless, the capsule is not flattened so that it still appears approximately spherical (Fig. 6). Therefore the flattening of discharged nematocysts should be ascribed to a change in the capsule wall which renders it less rigid.

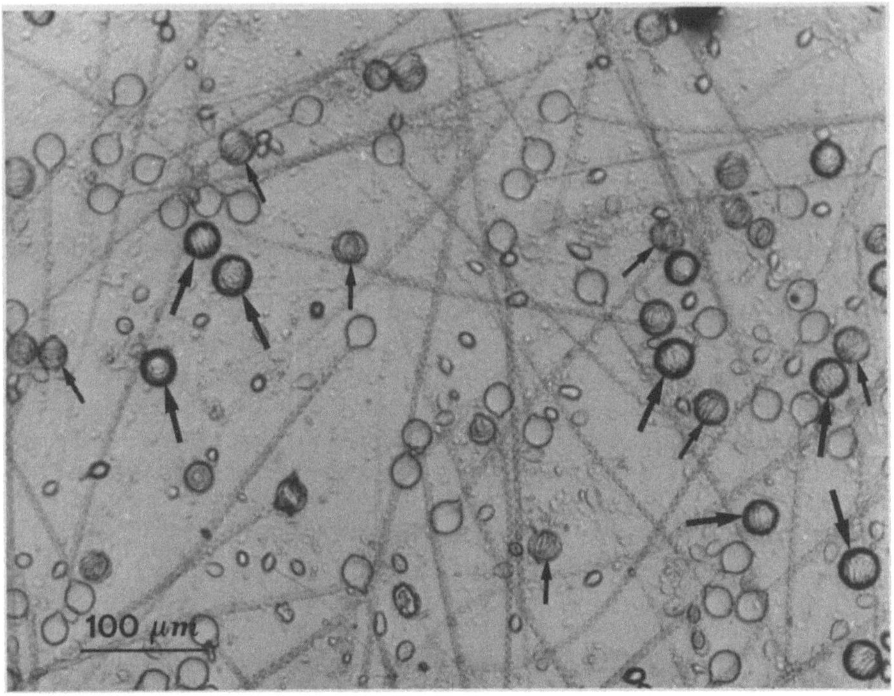

Fig. 2. Resting and discharged nematocysts. The discharged capsules are smaller than the resting ones. Among the undischarged capsules two distinct features can be observed: normal capsules *(large arrows)* and capsules that have permanently lost their discharging capacity *(small arrows)*

Fig; 3. S.E.M. picture of a normal resting nematocyst. Note the rough surface

Fig. 4. S.E.M. picture of a discharged capsule. The outer surface is similar to that of a resting capsule

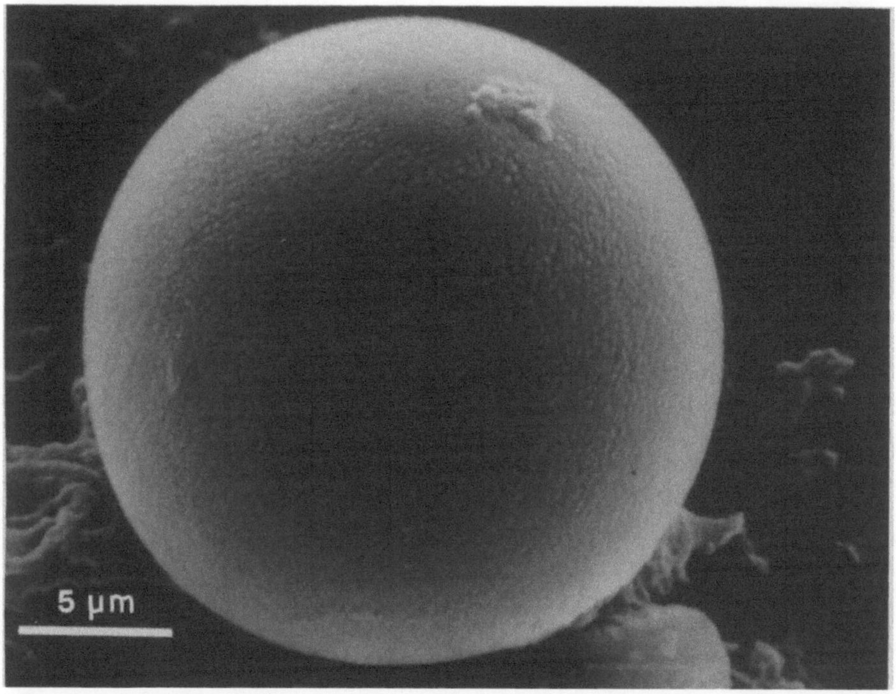

Fig. 3

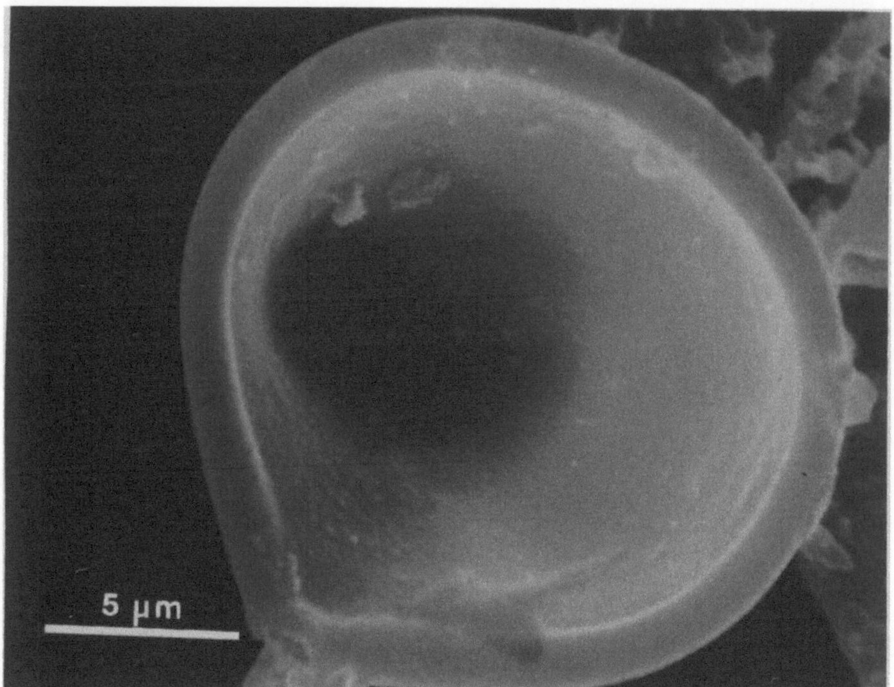

Fig. 4

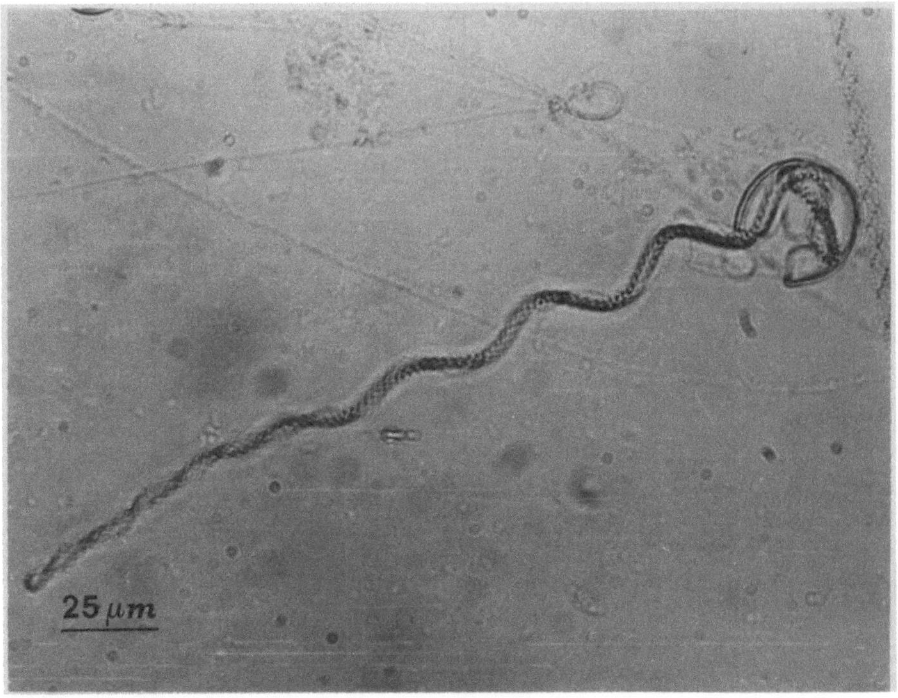

Fig. 5. Light microscope picture of a broken nematocyst

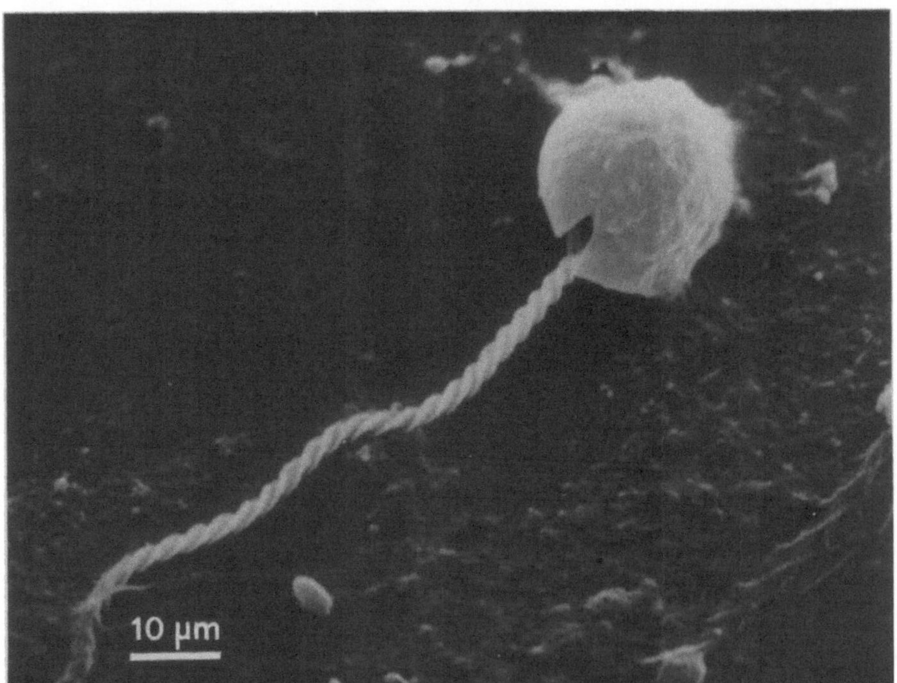

Fig. 6. A broken capsule at S.E.M. Note that the wall has not flattened

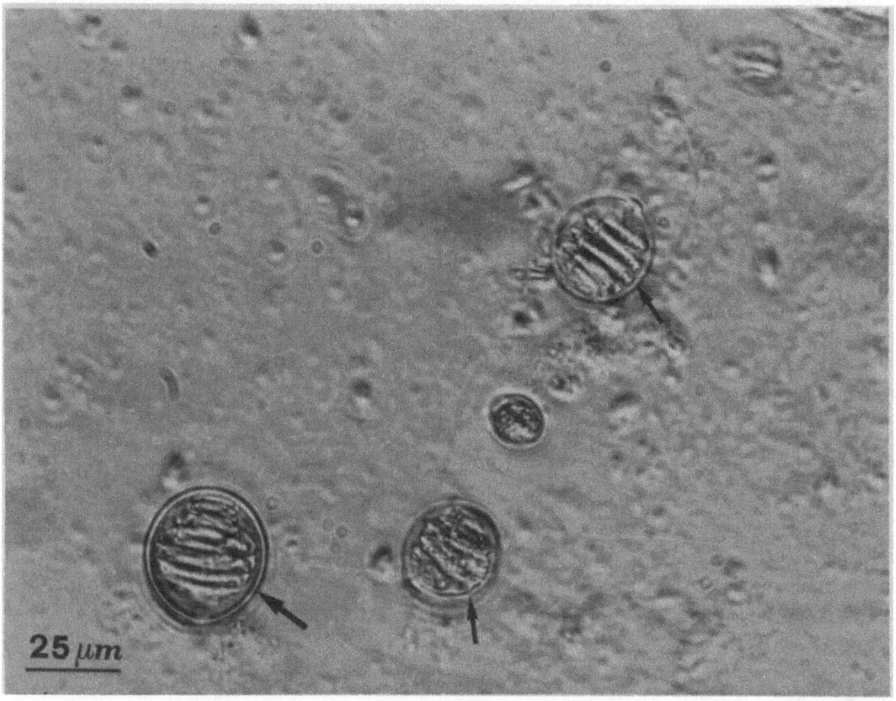

Fig. 7. A normal resting nematocyst *(large arrow)* and two collapsed capsules *(small arrows)*. Note that in the collapsed capsules the outer dark layer is lacking

The collapsed capsules can be easily distinguished from the normal ones even under a light microscope, since their wall is less marked (Fig. 7). At S.E.M. observation (Fig. 8) the capsule wall, apart from being adherent to the inner content of the nematocyst, which becomes clearly visible, is smooth when compared to that of resting (Fig. 9) and discharged nematocysts. The interpretation that can be given is that the outher rough layer of the wall has either dissolved or detached leaving a thinner and less rigid capsule wall. This change, which has been previously described as happening spontaneously at a slower rate in distilled water at room temperature, is associated with a permanent impairement of the discharge capacity (Salleo et al. 1983 b).

3.2 Permeability of Collapsed Capsules

The results of the test of permeability to dyes performed on normal and collapsed capsules are shown in Table 1. The normal capsules were freely permeable up to a molecular weight of 407.99. Therefore they are slightly less permeable than those of *Rhodactis rhodostoma* tested by Lubbock and Amos (1981). On the other hand, the collapsed capsules were permeable to all the dyes employed.

This could be due to large lesions of the wall produced by pH of the medium. Nevertheless, since collapsed capsules were not found to be permeable to human hemoglobin, such lesions should be excluded. Therefore, such increase in permeability

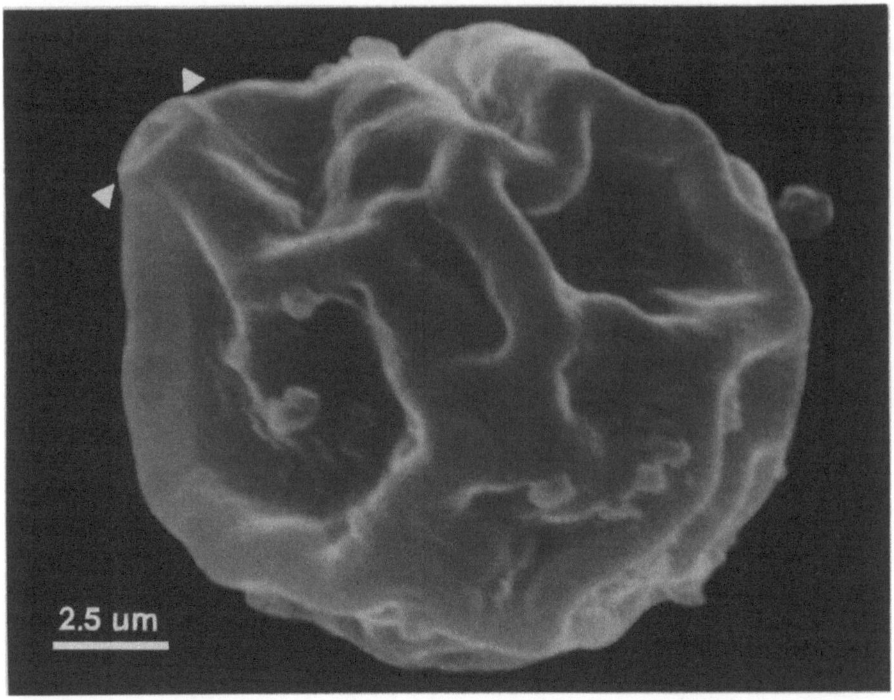

Fig. 8

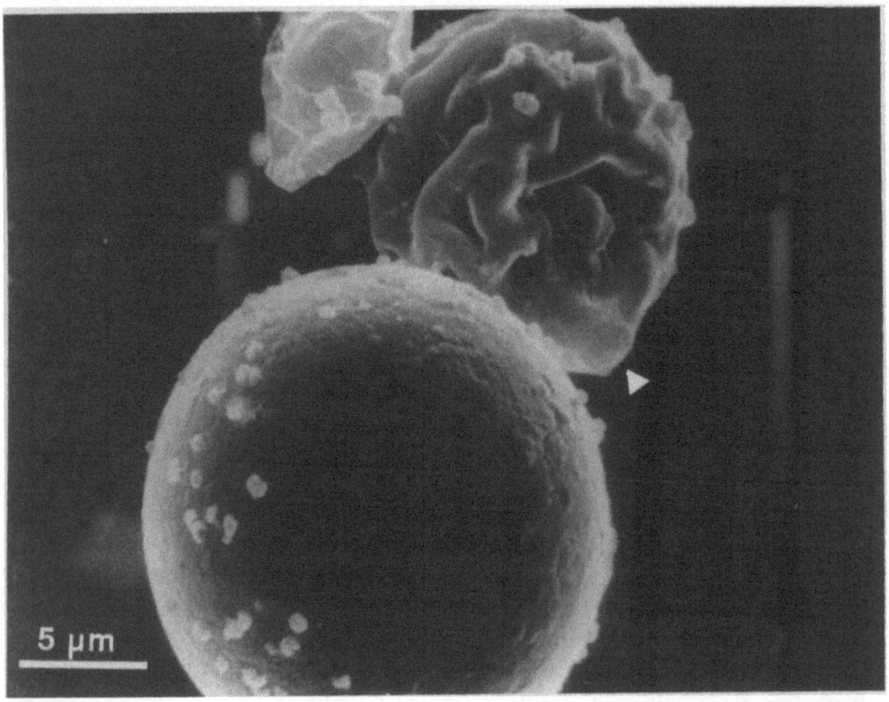

Fig. 9

Table 1. Permeability to dyes of normal and pH treated capsules

Dye	M.W.	Permeability	
		Normal nematocysts	Collapsed nematocysts
Neutral red	288.79	+	+
Methylene blue	355.90	+	+
Gentian violet	407.99	+	+
Janus green	511.09	–	+
Ruthenium red	551.22	–	+
Congo red	696.68	–	+
Aniline blue	737.70	–	+
Trypan blue	960.80	–	+
Alcian blue	1300	–	+
Hemoglobin	64,000	–	–

depends on the removal of the outer rough layer which exposes the underlying undamaged layer. The latter should have a chemical composition completely different from the outer one. In fact, once the outer layer is removed the wall becomes Gram-positive, while in normal resting capsules as well as in discharged ones it is negative. This does not mean that the smooth layer exposed by pH treatment has a chemical composition similar to the wall of bacteria, since it was not dissolved by lysozyme like the outer layer of resting and discharged capsules.

3.3 Effect of Polyvalent Cation

The effect of a previous treatment with $ThCl_4$ on I^- induced discharge is shown in Fig. 10. 20 mM $ThCl_4$ inhibited completely the discharging effect of I^-, and even at a concentration of 1 mM a significant, although not complete, inhibition was observed. Since $ThCl_4$ was removed from the medium prior to apply I^-, it should be inferred that Th^{4+} acts by binding to negatively charged groups placed either into the capsule fluid or on the capsule wall. The latter site of action seems to be more probable, since Ruthenium red, a dye which cannot penetrate through the wall (Fig. 10), produced an inhibition slightly lower than that of 10 mM $ThCl_4$.

On the other hand, neither Th^{4+} nor Ruthenium red inhibited the discharge induced by trypsin.

Fig. 8. A pH treated capsule. Note the smooth outer surface. The nature of the pillars connected to the operculum *(arrowheads)* cannot be specified

Fig. 9. A normal resting nematocyst *(front)* and a collapsed one *(back)*. Note the rough outer surface of the former. *Arrowhead* operculum

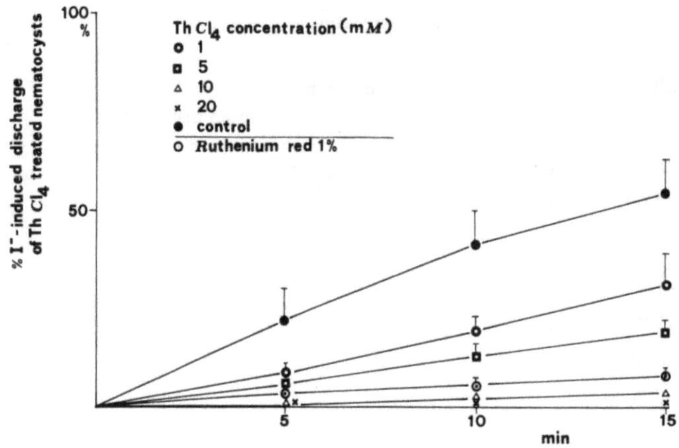

Fig. 10. Discharging rate following treatment with $ThCl_4$ and Ruthenium red

4 Discussion

The results of the present investigation reveal that some changes in the capsule wall should be associated with the discharge process.

The nematocysts of *Pelagia noctiluca* can be isolated in the resting state by treating the tentacle with distilled water. This means that they are insensitive to a very high osmotic gradient. Since an impermeability of the capsule wall to water should be excluded on the basis of its permeability to dyes up to a molecular weight of 407.99, the resistance to osmotic stress of resting capsules could depend on the rigidity of the capsule wall. This interpretation is confirmed by the lack of deformation in capsules submitted to dehydration. The comparison of resting normal capsules to broken ones ensure that even if the capsule lacks its contents, the approximately spherical shape is maintained.

The discharged capsules are smaller than the resting ones, although they maintain, in a suspending medium, an approximately spherical shape. This result contrasts with other observations, cited by Mariscal (1974), that some capsules enlarge after discharge. On the other hand, under dehydration, the discharged capsules tend to flatten. These results suggest that during the discharge process shrinkage and softening of the capsule wall take place. The S.E.M. picture does not show any difference between the resting and the discharged capsules as regards the rough outer layer.

The integrity of the latter seems to be of the utmost importance for the discharge. In fact, the results of present investigation show that when the outer layer is lacking, which happens spontaneously at room temperature in a few hours, as well as, at a higher speed, by treatment with pH lower than 3, the discharging capacity is permaently lost. The underlying smooth and thin layer is characterized by a higher permeability with respect to the outer one.

The results obtained with $ThCl_4$ and Ruthenium red show that the inhibitory effect of Ca^{2+} is not a specific one. Both Th^{4+} and Ruthenium red have been shown to be powerful inhibitors of the discharge induced by univalent anions. Moreover, they act by binding to some negatively charged molecules.

When the discharging agents effective on isolated nematocysts are taken into account, a conformational change of a protein molecule seems to be the main event in the discharge process, as suggested by Blanquet (1970). In fact the effectiveness of iodide (Salleo et al. 1983 a), an ion which has been employed by Szent-Gyorgyi (1951) for depolimerizing F actin to G actin, that of proteolytic enzymes (Yanagita and Wada 1954; Salleo et al. 1983) and that of pH (Yanagita and Wada 1953; Salleo et al. 1983 b) as well as the inhibitory effect of di-and polyvalent cations, correlate well with this hypothesis. Nevertheless, it is still uncertain whether the protein molecule involved in the discharge is placed into the capsule fluid, as suggested by Lubbock and Amos (1981) and by Lubbock et al. (1981) or on the capsule wall.

The latter possibility seems to be more liable, at least for the nematocysts of *Pelagia noctiluca,* since the discharging effect of trypsin and the inhibitory effect of Ruthenium red cannot be exerted into the capsule fluid, since they cannot penetrate through the capsule wall owing to their molecular weight. The softening of the capsule wall during discharge, which has been observed in the present investigation confirms this interpretation.

The conformational change of a protein molecule placed on the capsule wall should not induce the discharge through an osmotic mechanism, since the permeability of the capsule is high even in the resting state. It is proposed that it could act by changing the mechanical characteristics of the system.

The following model fits well with the following observations: (a) The undischarged capsule wall is rigid. (b) Shrinkage and softenting of the capsule wall are associated with the discharge. (c) The presence of the outer layer is essential for the discharge capacity. (d) The discharge can be elicited by molecules that cannot penetrate through the wall, such as proteolytic enzymes, and it can be prevented by molecules that cannot penetrate, such as Ruthenium red.

An elastic component of the capsule wall, maintained under a constant stretch could be counteracted by a rigid component, so that its elastic force cannot be applied to the capsule content. A conformational change of the protein responsible for the rigidity of the capsule wall could reduce this property. For instance, the collagenous-like protein which has been described in the wall of nematocysts (Blanquet and Lenhoff 1966; Mariscal and Lenhoff 1969) could be involved. A recent study (Svendsen et al. 1983) on the effect of pH on the mechanical characteristics of rat tail collagen, has shown that at pH lower than 3 a critical decrease in stiffness takes place.

On the other hand, it cannot be excluded that the rigid component could be placed in the capsule fluid, where a high concentration of a protein molecule mainly composed with glutamic acid has been described (Blanquet 1968; Calabrese et al. 1983). A change from α helix to random coil could reduce the mechanical resistance of the capsule content, so allowing the shrinkage of the wall and, consequently, the eversion of the thread.

Further experiments are needed to confirm the above model. Therefore, the mechanism of discharge of the nematocysts is still an unresolved problem.

References

Blanquet R (1968) Properties and composition of the nematocysts toxin of the sea anemone *Aiptasia pallida*. Comp Biochem Physiol 25: 893–902

Blanquet R (1970) Ionic effects on discharge of the isolated and in situ nematocysts of the sea anemone *Aiptasia pallida:* a possible role of calcium. Comp Biochem Physiol 35: 451–461

Blanquet R, Lenhoff HM (1966) A disulfide-linked collagenous protein of nematocyst capsules. Science (Wash DC) 154: 152–153

Calabrese L, Salleo A, Alfa M, La Spada G (1983) The release of a glutamate rich protein from *Pelagia noctiluca* nematocysts in the discharge process. 5th Conference of the European Society for Comparative Physiology and Biochemistry. Taormina (Italy) September, pp 5–8, 53–54

Cignitti M, Figura M, Marchetti M, Salleo A (1970) Electrokinetic effect in mechanoelectrical phenomenology of the bone. Arch Fisiol 68: 232–249

Cormier SM Hessinger DA (1980) Cnidocil apparatus: sensory receptor of *Physalia* nematocytes. J Ultrastruct Res 72: 13–19

Diamond JM, Wright EM (1969) Biological membranes: the physical basis of ion and nonelectrolyte selectivity. Ann Rev Physiol 31: 581–646

Ertman SC, Davenport D (1981) Tentacular nematocyte discharge and "self-recognition" in *Anthopleura elegantissima* (Brandt). Biol Bull (Woods Hole) 161: 366–370

Hamon M (1955) Cytochemical research on coelenterate nematocysts. Nature (Lond) 4477: 357

Holstein T (1981) The morphogenesis of nematocytes in *Hydra* and *Forskalia*: an ultrastructural study. J Ultrastruct Res 75: 276–290

Lubbock R (1979) Chemical recognition and nematocyte excitation in a sea anemone. J Exp Biol 83: 283–292

Lubbock R, Amos WB (1981) Removal of bound calcium from nematocysts causes discharge. Nature (Lond) 290: 500–501

Lubbock R, Gupta BL, Hall TA (1981) Novel role of calcium in exocytosis: mechanism of nematocysts discharge as shown by X-ray microanalysis. Proc Natl Acad Sci USA 78: 3624–3628

Mariscal RN (1974) Nematocysts. In: Muscatine L and Lenhoff HM (eds) Coelenterate biology. Academic, London, p 129

Mariscal RN (1980) The elemental composition of nematocysts as determined by X-ray microanalysis. In: Tardent P and Tardent R (eds) Developmental and cellular biology of coelenterates. Elsevier/North Holland Biomedical, p 337

Mariscal RN, Lenhoff HM (1969) Effect of disulfide reducing agent on coelenterate nematocyst capsule. Experientia (Basel) 25: 330–331

Phillips JH (1956) Isolation of active nematocysts of *Metridium senile* and their chemical compostion. Nature (Lond) 178: 932

Salleo A, La Spada G, Alfa M (1983) Blockage of trypsin-induced discharge of nematocysts of *Pelagia noctiluca* by Ca^{2+}. Mol Physiol 3: 89–97

Salleo A, La Spada G, Falzea G, Denaro MG (to be published 1983a) Discharging effect of anions and inhibitory effect of divalent cations on isolated nematocysts of *Pelagia noctiluca.* Mol Physiol

Salleo A, La Spada G, Falzea G, Denaro MG (to be published 1983b) pH-induced collapse of the capsular wall in isolated nematocysts of *Pelagia noctiluca.* Cell Mol Biol

Svendsen KH, Thomson G, Wismer-Pedersen J (1983) Absolute values of lenght-tension relations from collagen fibers of different ages and at different pH values. Mol Physiol 3: 237–247

Szent-Gyorgyi A (1951) A new method for the preparation of actin. J Biol Chem 192: 361–369

Tardent P, Honneger T, Baenninger R (1980) About the function of stenotheles in *Hydra attenuata pall.* In: Tardent P and Tardent R (eds) Developmental and cellular biology of coelenterates. Elsevier/North Holland Biomedical, p 331

Wood RL, Novak P (1982) The anchoring of nematocysts and nematocystes in the tentacles of *Hydra*. J Ultrastruct Res 81: 104–116

Yanagita TM, Wada T (1953) Discharge-inducing concentration of acids and bases for the nematocysts of sea anemone. Nat Sci Rep Ochanomizu Univ 4: 112–118

Yanagita TM, Wada T (1954) Effects of trypsin and thioglycollate upon the nematocysts of the sea anemone. Nature (Lond) 4395: 171

Role of High Concentrations of Ca, Cu, and Zn in the Maturation and Discharge in Situ of Sea Anemone Nematocysts as Shown by X-Ray Microanalysis of Cryosections[1]

B. L. GUPTA and T. A. HALL[2]

1 Nematocysts-an Overview

Animals in the Class Cnidaria (jelly fish, sea anemones etc.) produce a variety of toxins and other pharmacologically active chemicals (Lane 1968; Mariscal 1974) which are delivered into the victim by the stinging organelles called cnidae or nematocysts. They come in a plethora of forms and complexity of structures (Weill 1930; Mariscal 1974; Tardent and Tardent 1980). In essence, each one consists of a closed capsule containing an inverted harpoon-like barbed thread or filament and filled with chemically complex fluid including the toxins. Each nematocyst is contained within an ectodermal cell called a nematocyte. When stimulated the nematocyst can explosively discharge its thread. In some cases the whole nematocyst is fired out of the ectoderm during the discharging process (for example stenoteles of *Hydra*; Tardent and Tardent 1980). At one time it was thought that the nematocyst was an independent effector and therefore possessed a complete mechanism of receptor-response coupling not controlled in any way by the animal. Compelling evidence has now accumulated to discount this belief and show that the discharge of the nematocyst need not be involuntary but can be regulated either by the whole organism or locally by the epidermal tissue (Mariscal 1974; Conklin and Mariscal 1976). The complete mechanism of nematocyst discharge must therefore involve signal recognition, transmission, transduction and response.

In some cases the signal has to be highly specific. An interesting example of this specificity is the colonal *Anthopleaura elegantissima* (Lubbock 1980). This anemone lives in cloned colonies (Fig. 1A) and in addition to normal feeding tentacles with their cnidae, possesses specialised organs called acrorhagi (Fig. 1B) packed with large nematocysts (Fig. 2A). The general behaviour of anthozoans leading to the discharge of oral and acrorhagial nematocysts has been investigated by a number of previous workers (see Mariscal 1974) and in elegant detail by Lubbock (1980). The oral tentacles involved in the capture of food discharge their nematocysts on contact

---

1 This paper is dedicated to the memory of Dr. Roger Lubbock who died in a tragic automobile accident in Brazil on 5 September 1981 at the age of 29 years. The experimental data discussed in this paper are based on the work, both published and previously unpublished, which was carried out in collaboration with Dr. Lubbock. The authors are, however, responsible for the contents of the present article.

2 Department of Zoology, Cambridge University, Downing Street, Cambridge CB2 3EJ, England

Toxins, Drugs, and Pollutants in Marine Animals
(ed. by Bolis et al.)

with the external surface of a whole range of metazoan animals from Porifera to Vertebrata, and occasionally even with inanimate objects. However, the contact of oral tentacles with other anthozoan species usually induced a distinct "aggressive" behaviour (strictly speaking a defensive behaviour, Marsical 1974), leading to the extension of acrorhagi (Fig. 1B). The discharge of acrorhagial nematocysts occureed only when the surface of the ectodermal cells in acrorhagi was brought into physical contact with the cell surface (not the mucus layer) of the other Anthozoa. Contact of acrorhagi with the members of other allogeneic clones (genetic cousins) induced a particularly violent response leading to a simultaneous and massive firing of nematocysts (Fig. 2B, C). The response was confined to the acrorhagus contacted but frequently extended 1–2 mm from the point of contact suggesting a cell-to-cell transmission of the signal. Neither the oral tentacles nor the extended acrorhagi responded to a contact with syngeneic clone mates.

From these studies on the aggressive behaviour the following conclusions were derived by Lubbock (1980). (1) The ectodernal cells in oral tentacles of *A. elegantissima* possess a population of chemoreceptors similar in specificity to the receptors on acrorhagi but presumably different from the general receptors. (2) These oral tentacle receptors are stimulated by all foreign Anthozoa leading to the aggressive behaviour (protrusion of acrorhagi). (3) Acrorhagi have highly specific chemoreceptors, presumably situated on the ciliary cones (Mariscal et al. 1978; Lubbock and Shelton

Fig. 1 A, B. ***Anthopleura elegantissima.*** **A** Two allogeneic clones in the Laboratory aquarium: in life the top clone had a deep green and the bottom clone a steel-blue pigmentation. **B** Two allogeneic individuals showing long and slender oral tentacles and short and stubby acrorhagi **(rocket arrows)** in an excited state. (Photographs by (late) Dr. Roger Lubbock)

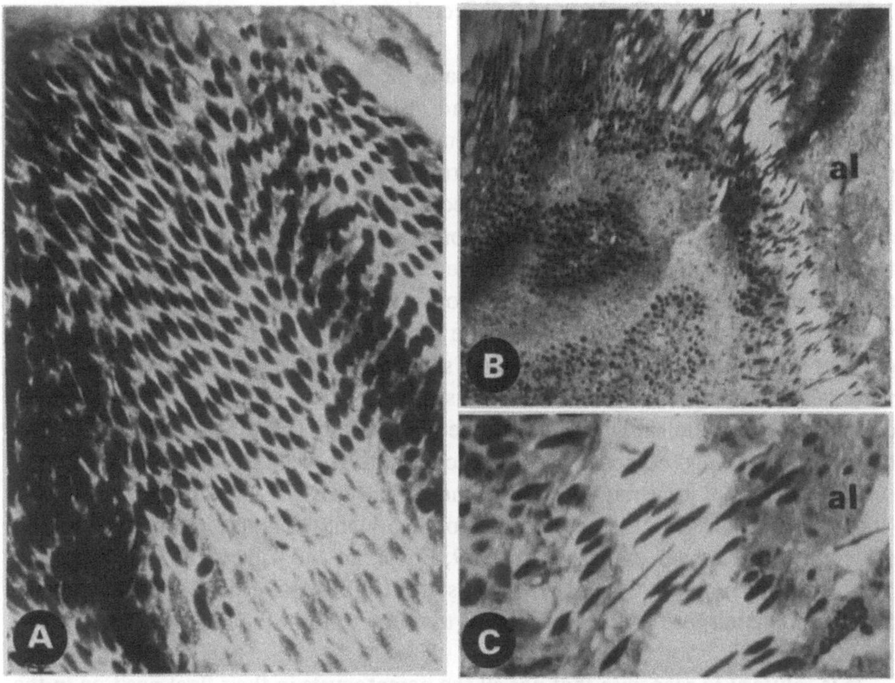

Fig. 2 A–C. ***Anthopleura elegantissima:*** **quench-frozen Acrorhagi. STEM images of unstained frozen-dried cryosections to show the nematocysts (A) in the resting state and (B, C) after the discharge had been induced with a bit of allogeneic** ***(al)*** **column tissue**

1981) which are only excited by contact with foreign Anthozoa and particularly vigorously, with allogeneic individuals, causing the discharge of nematocysts. (4) It seems that the acrorhagial receptors respond to the surface substances bound to the plasma membranes of the ectodermal cells of foreign anemones ("altered self") in the sense of vertebrate histocompatibility restriction. (5) The colonal recognition phenomenon appears to be genetically determined and is not a consequence of habituation. It would appear therefore that the phenomenon of nematocyst discharge is very much like the stimulus-response coupling in cell secretion for which a unifying hypothesis is being sought (Rasmussen 1981; Robin 1982). Since a contact between the plasma membrane surfaces appears to be required for an excitatory response the phenomenon is analagous to sperm-egg interaction in fertilization (Epel 1977, 1980) but is even more specific because the acrorhagial receptors do not respond to contacts with syngeneic cell surfaces.

1.1 Nematocysts as the Products of Cell Secretion

Nematocysts (and other cnidae) are perhaps the most complex structures synthesized, stored in fully differentiated state and discharged in response to a stimulus by the terminally differentiated cells. The nematoblast located basally in an interstitial pocket in the ectoderm, when induced, grows and differentiates to become a polarised nematocyte with external surface specialisations such as microvilli, cilia, ciliary cones, etc. and the basal surface resting on a basal (basement) lamina. At the same time the nematocyst material is synthesized in rough endoplasmic reticulum (RER) and Golgi complexes to form a large ovoid secretion vesicle – the capsule – continuous with a "tube", often coiled – the filament. In transmission electron micrographs (TEM) of thin, heavy-metal stained sections of conventionally fixed tissue, the material of the capsule wall during early secretion is not stained while the granular contants of both the capsule and the tube are stained. The tube appears to be "modelled" by the Golgi complexes which encircle the growing tip of the tube and presumably add to it secretion vesicles. As the nematoblast migrates to the surface, the "tube" invaginates into the capsule and forms a highly complex pleated and barbed structure (Slautterback and Fawcett 1959; Chapman and Tilney 1959; Slautterback 1961; Chapman 1961; Skaer 1973; Holstein 1980; Germer and Hundgen 1980). Some workers dispute the invagination of the tube to form the filament (Germer and Hundgen 1980).

It is salutary to recognise that not all the nematocysts in their final position in the ectoderm are fully mature. For example Skaer (1973) emphasised that in conventional TEM's of the immature nematocysts of the siphonophoran *Rosacea cymbiformis,* even in the final position, the capsular fluid stained darkly with heavy metal stains while the thread wall was pale. In fully mature nematocysts the capsular fluid was unstained but the filament wall appears very electron dense. Some earlier histochemical work with the light microscope had suggested that in the course of maturation there is a decrease in negative surface charge of the nematocyst contents (see Picken and Skaer 1966). These observations would appear to be relevant to the changes in the elemental composition of the nematocyst during maturation discussed in Section 2.2. These observations are also relevant to the confusion which exists in the literature on the discharge of nematocysts under various experimental conditions (Mariscal 1974; Salleo 1983). Some confusion also exists with regard to the final fate of the cell membrane enclosing the secretion vacuole into which the nematocyst is formed (Robson 1973). It is clear in the classical TEM work on *Hydra* that during all stages of formation and in fully matured form the nematocyst is enclosed within a continuous cell membrane (Slautterback 1961; Chapman 1961). The situation in marine organisms with large nematocysts seems less clear. Robson (1973) noted that in the anthozoan *Gonactinia* a cell membrane is always evident over the front end of the spirocyst, underneath the apical plasma membrane, but the basal part of the capsule seems to lack a continous cell membrane. Instead the capsule wall is in contact with tubular ER. In the TEM of the nematocysts from acrorhagial tissue of *A. elegantissima* fixed with glutaraldehyde/tannic acid the cell membrane of the vacuole enclosing the capsule appears to form numerous, convoluted tubular extensions frequently in contact with capsule wall (Gupta, Lubbock and Bray, unpublished results). These tubular extensions of the vacuole membrane might have been mistaken for tubular ER by Robson (1973). However, her suggestion that these tubular elements have an important role in regulating the inoic changes during both maturation and discharge is highly plausible (see Sect. 2.2.5).

In summary, the nematoblast secretes the molecular components of the nematocyst first as a featureless mass of macromolecules into a membrane-bound secretion vesicle. This material must contain at least: (1) the glutamic and aspartic acid rich peptides as well as toxins of the capsule fluid (Lane 1968; Picken and Skaer 1966; Mariscal 1974); (2) the collagenous material for the fibrous double-layered capsule wall (Lenhoff et al. 1957; Picken and Skaer 1966; Robson 1973; Mariscal and Mclean 1976); (3) sulphur-rich metalloproteins for the hollow filament with ‚information' for forming complex pleats, folds, etc. (Skaer and Picken 1965; Skaer 1973; Marsical and Mclean 1976); (4) proteins to form barbs, spines, harpoon, etc. shaped and arranged in a characteristic fashin for each type of cnidae in every species of Cnidaria (Marsical 1974); and finally (5) the material to form a hinged flap or operculum weakly bonded to the capsule wall. Temporal changes in the spatial organisation of RER-Golgi in the cytoplasm of the cnidoblast recorded by several electron microscopists suggest a programmed secretion of different molecular constituents but the final differentiation and packing of the nematocyst structure appears to result from self-assembly within the capsule. What regulates this process of self-assembly has remained a fascinating mystery.

1.2 Mechanism of Nematocyst Discharge

The mechanism which causes a nematocyst (and other cnidae) to respond to a signal, unfold its filament and discharge with force has occupied biologists' attention for some 200 years. Numerous hypotheses have been proposed and the subject has been extensively reviewed (Picken and Skaer 1966; Mariscal 1974; Robson 1973). Although much of the early work was done by studying the mechanism in situ, more recent studies have been carried out on large nematocysts freed from the cells. These studies on isolated nematocysts have provided some valuable insights into the mechanism by ruling out many previous suggestions but their relevance to the actual mechanism in situ must be considered with caution. Blanquet (1970) correctly emphasised that a great many experiments on isolated nematocysts have been performed with harsh treatments not likely to be encountered within the nematocyte in situ. Marsical (1974) has also pointed out that in several experiments the term "discharge" has been loosely applied to mean sometimes the expulsion of the nematocyst from the tissue and at other times the eversion of the filament. Here we use "discharge" to mean filament eversion. Any plausible hypothesis for a common mechanism of cnidae expulsion and discharge must accommodate at least the following facts. (1) In organisms such as *Hydra* the discharge of cnidae in situ occurs within a few milliseconds of stimulation. (2) The electrical stimulation of tissue bearing large nematocysts in *Corynactis* can bring about discharge within a fraction of a second (Picken and Skaer 1966). (3) Massive discharge and expulsion of nematocysts after touching the tip of acrorhagus of *A. elegantissima* with a bit of allogeneic tissue is virtually instantaneous (Lubbock and Shelton 1981). Acrorhagi mounted on the tips of copper pins, stimulated similarly and quench-frozen within 1/2 s for microanalysis had masses of discharged nematocysts shot into the allogeneic tissue (Fig. 2B, C). The control tissue simiarly mounted but frozen after touching with a glass rod had no discharged nematocysts. (4) Robson (1953) observed that nematocysts of *Corynactis* will discharge even after prolonged heat treatment. These observations would rule out those hypotheses which require enzymatic or other relatively slow structural modifications of the nematocyst components.

In essence, a rapid discharge of cnidae could involve at least 3 mechanisms which might not be mutually exclusive.

Contractile Mechanisms. The force with which stenoteles are fired as 'darts' and 'missiles' into the target (Tardent and Tardent 1980) suggests that a contractile mechanism within the cnidocyte and/or the neighbouring columnar (myoepithelial) cells is involved in the ejection process (Chun 1881; Slautterback 1967). This possibility has been generally dismissed because contractile structures could not be found in all cases (Robson 1973). Now it is accepted that virtually all cells have acto-myosin type and other cytoskeletal structures which are capable of contraction (Alberts et al. 1983). However, such contractile systems are not likely to be the primary mechanisms in the actual filament discharge because it can be achieved in isolated cnidae even under realistically mild conditions without any cell components. The role of such contractile systems either within the cnidocyte or collectively in the myoepithelium in inducing forceful firing (ejection) of cnidae that occurs in many species needs further investigation.

Changes in the Mechanical and/or Permeability Properties of the Capsule Wall. No evidence has yet been found that the collagenous capsule wall is contractile: nor would one expect active contractile properties from collagen (Robson 1973). However Robson did find that in the isolated holotrichous isorhiza nematocysts of *Corynactis* and *Rhodactis* at the initiation of discharge the capsule may increase in volume to 150% of the resting state. After the eversion of the filament the capsule first decreased in volume rapidly, attributable to the loss of some 50% of the capsular volume previously occupied by the filament. Subsequently the capsule returned exactly to its resting volume and therefore had a net gain in fluid volume over the undischarged state. These observations by Robson support the suggestions of earlier workers that the capsule wall is elastic but not the suggestions that it is under tension in the resting state.

Salleo (1983) working on the isolated holotrichous isorhiza nematocysts of the jelly fish *Pelagia noctiluca* has suggested that the capsule wall of the undischarged nematocysts is rigid and therefore responsible for maintaining a constant shape and size even when dehydrated. The discharge-inducing mechanism somehow causes the capsule wall to weaken or soften. The discharge is then effected by the elastic component of the wall exerting pressure on the contents. A 40% loss in the capsular volume after discharge can be accounted for by the loss of filament (Robson 1973) but Salleo (1983) believes that the reduction in volume is due to the elastic collapse of wall.

It should be noted that Salleo measured the volume changes in isolated nematocysts under phase-contrast optics where the apparent diameters of spherical objetcts

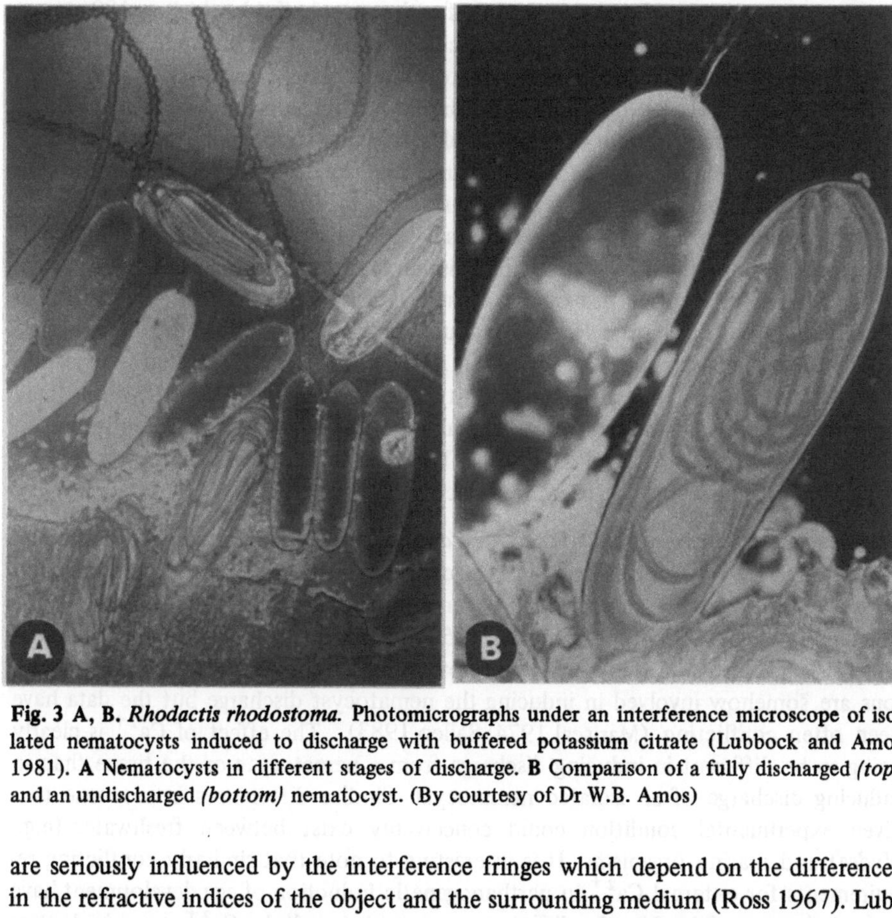

Fig. 3 A, B. ***Rhodactis rhodostoma.*** **Photomicrographs under an interference microscope of isolated nematocysts induced to discharge with buffered potassium citrate (Lubbock and Amos 1981). A Nematocysts in different stages of discharge. B Comparison of a fully discharged** ***(top)*** **and an undischarged** ***(bottom)*** **nematocyst. (By courtesy of Dr W.B. Amos)**

are seriously influenced by the interference fringes which depend on the differences in the refractive indices of the object and the surrounding medium (Ross 1967). Lubbock and Amos (1981) found by interference microscopy that the capsules of undischarged nematocysts in *Rhodactis* have a refractive index of about 1.385 but the final index of fully discharged capsules was only about 0.56% higher than that of the surrounding sea water (Fig. 3A, B). It is not clear how Salleo in his phase-contrast measurements of diameters compensates for such gross differences in refractive indices. Similarly Salleo has used fixation, liquid dehydration and critical point drying to prepare samples for SEM. These methods of preparation can cause gross distortion and collapse of delicate objects (Beckett et al. 1982).

In any case, the observation by several previous workers (Robson 1973; Mariscal 1974) that the capsule expands briefly before discharge suggests an initial increase in the hydrostatic pressure caused by the capsule fluid rather than the inverse suggested by Salleo.

Picken and Skaer (1966) first made the observation that the fluid collected from the undischarged nematocysts isolated from *Corynactis* had a freezing point between -4.16° and -5.94°C as compared with -1.875°C for the sea water used and -3.435°C for the standard molar NaCl solution. The capsular fluid at the end of discharge had a freezing point of -3.83°C. Lubbock and Amos (1981) repeated these observations on *Rhodactis* using intact but undischarged nematocysts in situ and found freezing point drpession values of 5.4° to 8.2°C ($\bar{x} = 6.2$, n = 15). These values correspond to a *notional* osmotic pressure for the capsule fluid as high as 180 atmospheres. However the capsule in the undischarged state is not a simple salt solution but contains macromolecules making up to some 40% dry weight (Blanquet 1968, 1970; Lubbock et al. 1981). There is no simple relationship between freezing point depressions of concentrated macromolecular solutions and their osmotic pressure (Wolf 1966) and therefore one cannot assume that the effective osmotic pressure inside the undischarged nematocysts is constantly high.

Nevertheless it has been proposed that such may be the case and that undischarged capsule is impermeable to water. The discharge is then made possible by a change in the permeability properties of the capsule (Robson 1973). This possibility can now be ruled out because in *Rhodactis* (Lubbock and Amos 1981) the undischarged capsule is permeable to substances ofup to M.W. 670 and in *Pelagia* (Salleo 1983) up to M.W. 407. In spite of this high permeability, the undischarged nematocysts isolated from these marine animals, do not generally discharge in sea water, confirming an earlier observation by Yanagita (1943). Therefore the contents of the resting undischarged capsule cannot be osmotically active.

Changes in the Osmotic Properties of the Capsule Fluid. Numerous workers have previously suggested that an osmotic uptake of water causes a rise in the intracapsular pressure leading to the removal of the operculum and the discharge (eversion) of the filament (Picken 1953, 1957; Robson 1953, 1973; Slautterback 1961; Picken and Skaer 1966; Skaer 1973, etc.). Several previous studies have indicated that calcium ions are somehow involved in inducing the nematocyst discharge but the data have been often conflicting (Mariscal 1974; Salleo 1983). The effect of Ca^{2+} is clearly goihg to be different in inducing discharge from a nematocyte or the tissue than in inducing discharge of an isolated nematocyst. Similar differences in response to a given experimental condition could conceivably exist between freshwater (e.g., *Hydra*) and marine organisms. It is interesting to note that similarly conflicting requirements for external Ca^{2+} in parthanogenetic induction of egg development have recently been explained by the different nature of intracellular Ca^{2+} stores in deutrostomic and protostomic eggs (Jaffe 1983).

Lubbock and Amos (1981) followed Blanquet's (1970) approach under more rigourously controlled conditions in the isolated holotrichous isorhiza nematocysts of *Rhodactis.* They discovered that a 10 mM solution of potassium citrate buffered at pH 7.0 with 10 mM imidazole induced nematocyst discharge. Citrate was thougt to act as chelating agent because EGTA instead of citrate also stimulated discharge. In each case 50 mM $CaCl_2$ in the solution inhibited citrate-(or EGTA)-induced discharge (cf. Blanquet 1970). The possibility that 50 mM $CaCl_2$ might render the capsule wall impermeable to water was ruled out by using dye markers: even in Ca-inhibited nematocysts the dye bromophenol blue (M.W. 670) penetrated the capsule without causing discharge. The citrate-induced discharge was not inhibited when in place of $CaCl_2$, 50 mM of NaCl, KCl or $MgCl_2$ were used. The interpretation therefore is that Ca^{2+} somehow is stabilising the internal macromolecular contents in undischarged nematocysts to render them osmotically inactive. Citrate ions (or other Ca chelating agents) have a higher binding constant for Ca^{2+} than the macromolecules in the capsule. Removal of Ca^{2+} brings the calcium binding macromolecules into solution thus raising the osmotic pressure of the capsular fluid and inducing discharge. However, a previous study by Mariscal (1980) on the X-ray microanalysis of nematocysts did not find Ca in any significant amount. The samples for this study were whole, isolated nematocysts, washed in distilled water for 15 min (Mariscal 1980) and air dried. The analysis therefore refers essentially to the structural components of discharged nematocysts (capsule wall, filament, spines and operculum) rather than the fluid contents which had been carefully washed away. In order to investigate the role of metal ions in the formation, maturation and discharge of nematocysts in situ Lubbock et al. 1981) employed the technique of examining 1–2 μm thick cryosections by electron probe X-ray microanalysis as developed and practised in the author's laboratory (Gupta 1976; Gupta and Hall 1978, 1981, 1982; Hall and Gupta 1983). The findings, previously published (Lubbock et al. 1981) as well as some unpublished data, are summarised below.

2 Electron Probe X-Ray Microanalysis (EPXMA)

2.1 Method

The mesenterial filaments of the corallimorpharian anemone *Rhodactis rhodostoma* and un-inflated acrorhagi from the sea anemone *Anthopleura elegantissima* were used. To prepare tissue for microanalysis, freshly excised filaments or acrorhagi were placed in artificial 21/25 sea water with 16.8% wt/wt dextran (Mγ 250,000) added to give a final composition in mmol kg^{-1}: Na, 328; Cl, 382; Mg, 38; SO_4, 20; K, 7; Ca, 7; HCO_3, 2; Br, Sr, B and F $<$ 1 mmol each. The tissue was mounted in a drop of dextran + sea water on the top of copper pins and quench frozen in Freon-13 (supercooled to about $-190^{\circ}C$) using a triggered mechanical device. To obtain samples of nematocysts in different stages of discharge the tissue mounted on copper pins was stimulated immediately before freezing. Discharge was induced by electrical stimulation in *Rhodactis* filaments and by touching the tip of an inflated acrorhagus with a small piece of column tissue from an allogeneic *A. elegantissima*. 1–2 μm thick cryosections of the frozen tissue were cut at about $-75^{\circ}C$ in a cryomicrotome and examined on the cold transmission stage ($-170^{\circ}C$) of a JEOL JXA–50A microprobe analyzer operated at 45 kV. The X-ray data were collected with two diffracting spectrometers (used for Na and Ca) and a Kevex-Link X-ray energy spectrometer with a Link Systems multichannel analyzer backed by a computer. Link Systems QUANTEM/FLS software was used for processing X-ray data, using continuum normalisation to provide instant print-out values of elemental concentration in mmol kg^{-1} of the specimen area under the beam. Sections were analysed initially in fully hydrated state to obtain elemental concentrations in mmol/kg wet wt and subsequently after dehydration within the microanalyser for values in mmol kg^{-1} dry wt. A comparison of the two concentration values permitted an exact estimation of water contents in the analysed micro-fields. Details of these procedures have been provided elsewhere (Gupta and Hall 1978, 1981, 1982).

It is most important to appreciate the following aspects of EPXMA. (1) The technique measures total elemental concentrations including components of organic

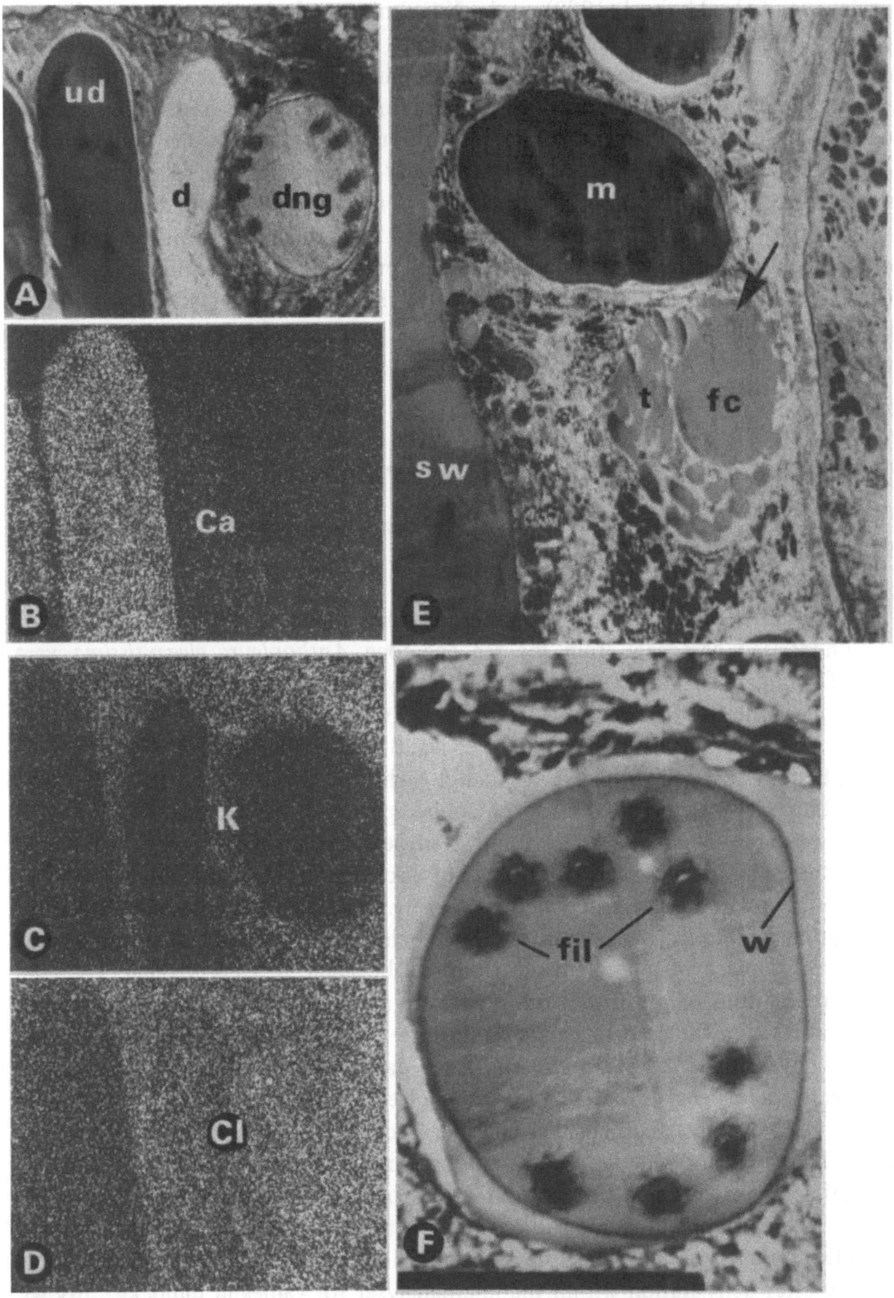

Fig. 4 A–F. *Rhodactis rhodostoma.* **A** STEM image of a frozen-dried cryosection showing the nematocysts arrested in three states: undischarged *(ud)*, discharging *(dng)* and fully discharged *(d)*. **B–D** Maps of X-ray signal *(white dots)* from a corresponding field in hydrated state; the spectrometer "windows" used for these maps include characteristic as well as continum X-ray quanta. **E** STEM image of a frozen-dried cryosection to show a mature *(m)* and a forming nematocyst with the capsule *(fc)* and a tube *(t)* arising from the capsule *(arrow)*. **F** STEM image of a

molecules (especially P and S because C, N, H and O are not measured), bound or salt-linked ions and ions in free solution. (2) In our method the local dry mass and therefore free water is also measured. (3) All analysable entities are measured *simultaneously* in each area of the section under the electron probe. (4) The probe in our system can be a highly focussed static beam of about 100 nm in diameter or it can be a scanning raster of any suitable dimensions (see example in Hall and Gupta 1983). (5) Even in 1–2 μm thick sections of nematocytes the spatial resolution of elemental analysis was found near 0.3 μm (Hall und Gupta 1983) However the image resolution in 1–2 μm thick sections is limited and does not permit identification of organelles (mitochondria, ER, Golgi). Due to the complexity of structure it was not therefore possible to resolve clearly the capsule wall or the components of highly folded filament from the capsule fluid in hydrated sections of undischarged nematocysts of *Rhodactis.* These structures are clearly identified and resolved in frozen dried sections (Fig. 4A, E, F; also see Lubbock et al. 1981; Gupta and Hall 1981). In *A. elegantissima* nematocysts (size about 2 X 50 μm) where the electron density of the undischarged capsules is very high, this level of structure and analytical resolution was often not possible.

2.2 Results

The X-ray analytical results from the nematocysts of *R. rhodostoma* are summarised in Table 1. These results have been simplified and re-arranged from the original data published by Lubbock et al. (1981) but include additional information on Ca, Cu and Zn in the filament and capsule wall. The standard errors (SE) of the mean values given by Lubbock et al. (1981) have been omitted for reasons of simplicity but were as a rule well within ± 10% of the mean values. Here we focus attention on some of the salient features of the role of ions in the assembly and discharge of the nematocysts.

Capsule Fluid. As discussed by Lubbock et al. (1981) the outstanding revelation of this study is the presence of 600 mmol kg^{-1} wet wt. of Ca in the capsule fluid of undischarged nematocysts. This level of Ca was not found anywhere else and it contrasts sharply with only 10–20 mmol kg^{-1} wet wt. total Ca found in the sea water, in the nematocyte cytoplasm, in the contents of immature nematocysts (Fig. 4E) or in the capsule wall and filament. The high values of Ca in the capsule wall and uneverted filament given by Lubbock et al. (1981) were based on the data from 2 μm thick, fully hydrated sections and reflect the high Ca in the fluid which permeates the inner folds of the capsule wall and the spaces between the intricate surface folds of the filament. When a highly focussed beam (≅ 100 nm in diameter) was localised on the dense parts of the capsule wall and the filamend in frozen-dried sections, the calcium levels corresponded to only about 20–40 mmol kg^{-1} wet wt.

nematocyst arrested in an early phase of discharge by cryofreezing, to show the differential contrast of the capsule wall *(w)* and the filament *(fil)*. Note that neither the tissue before freezing nor the cryosections were fixed or stained with any chemical solutions. The contrast in these STEM images (**A, E, F**) is due to intrinsic differences in the mean atomic number composition and the local dry mass fractions. *sw* is sea water + dextran

Table 1. Mean concentrations of elements (mmol kg^{-1} wet mass) and dry mass in *R. rhodostoma* nematocysts during formation after maturation and in fully discharged states. SE values given in the original data (Lubbock et al. 1981) have been omitted here

Component	Stage	Na	K	Cl	Ca	Mg	Cu[c]	Zn[c]	P	S	% dry mass
Cytoplasm[a]											
	Early nematoblast	43	108	106	9	32	nd[e]	nd	38	135	18
	mature nematocyte	16	100	99	9	32	nd	nd	30	121	16
	after stimulation	117	116	167	11	33	nd	nd	46	98	19
Capsule fluid											
	Early nematoblast	29	137	90	17	nd	nd	nd	17	156	18
	mature, undischarged	32	11	17	606	33	nd	nd	nd	54	32
	discharging	194	7	105	189	18	nd	<20	nd	18	22
	fully discharged	350	9	401	35	45	nd	nd	nd	37	11
Capsule wall											
	Mature, undischarged[b]	93	62	89	21	28	nd	nd	5	181	52
	discharging nematocyst	188	59	125	18	31	nd	nd	6	136	42
	fully discharged	140	61	181	15	29	nd	<10	5	162	50
Filament											
	Tube in nematoblast	39	105	147	8	7	nd	nd	49	118	19
	mature uneverted[b]	106	22	15	42	21	~30	~80	3	142	42
	fully everted	201	18	130	92	17	~30	~10	2	151	31
Vacuole[d]											
	Discharging nematocyst	234	11	389	531	24	nd	<20	2	18	19
21/25 sea water + dextran		328	7	382	7	38	nil	nil	0	20	19

[a] Cytoplasm contained numerous dense granules loaded with copper and zinc. These granules were excluded.
[b] Data are based on measurements which did not show high Ca given by Lubbock et al. (1981): the latter is considered as due to capsule fluid trapped in pockets (inner folds).
[c] These values are approximate; exact quantification was not done.
[d] Vacuole refers to the space between the capsule and the nematocyte membrane
[e] nd = detected

Surprisingly, the capsule fluid had very little chloride, no phosphorus and only about 50 mmol kg^{-1} wet wt. of sulphur. Sulphur is likely to be a part of the macromolecules constituting some 30% of the dry wt. (also compare Lubbock and Amos 1981). As noted above, the contents of the undischarged capsule appear to be osmotically inactive. This would appear to rule out the presence of bicarbonate or hydroxyl anions to maintain electroneutrality. The interpretation therefore is that the calcium in the undischarged capsule is not free but forms salt-links with the negatively charged polypeptides very rich in glutamic and aspartic acid (Blanquet 1968; Mariscal 1974). Both these acidic amino acids are characteristically high in other Ca-binding proteins in biological systems (Williams 1977).

It is noteworthy that the elemental compositon of the capsule forming in the nematoblast (Fig. 4E) is very similar to that of the surrounding cytoplasm. During the final maturation stages there must be a transport of Ca from the cell into the capsule across the membrane of the secretory vacuole together with the removal of potassium. These changes in the capsule fluid are likely to be the last step in the maturation of the nematocyst (see below) and would explain why the nematocysts which structurally look fully mature are otherwise unable to discharge (see Sect. 1.1).

Capsule Wall. The wall of the undischarged capsule had the highest concentration of sulphur (181 mmol kg^{-1} wet wt.) consonant with its sulphur-rich collagenous nature. In contrast to both the capsule fluid and the surrounding nematocyst cytoplasm, it also had high Na, K and Cl and 52% dry wt. The difference between the Na and K values in the capsule wall and those in other adjacent components was statistically significant. This would again suggest that these ions are mostly bound. Low levels of calcium (about 20 mmol kg^{-1} wet wt.) which remain more or less constant during the process of discharge together with its generally elastic properties (Robson 1973) would rule out any participation of the capsule wall in the calcium effect.

In transmission electron micrographs of concentionally prepared, resin embedded material the capsule wall in all the stages of differentiation appears as "unstained" or electron transparent (Slautterback 1961; Chapman 1961; Skaer 1973; Germer and Hundgen 1980). This is in sharp contrast with the STEM images of unifixed, unstained, cryosections where the wall of the immature capsule has the same electron density as the capsule fluid. This is not surprising, because in STEM images of cryosections the contrast is determined by the atomic number and total mass of the native structure: it does indicate that the wall material has very different ion binding properties from the fluid contents of the capsule in spite of a similar dry mass. During maturation the density of the capsule wall in STEM images is higher than that of the capsule fluid as is its dry mass. The changes in ionic environment during the final maturation stage are likely to be involved in the polymerization of procollagen into fibrous collagen of the wall.

Filament. The uneverted filament of the mature, undischarged capsule has, in addition to high sulphur, high concentrations of Zn, Cu and Na, but low K and Cl. Again this elemental compositon is significantly different from that of either the capsule wall, with which it is continuous at the base, or of the capsule fluid in which it is embedded. The exact quantification of Cu and Zn X-ray signals was not attempted but from relative sensitivities of the X-ray energy spectrometer to various X-ray lines, it is estimated at about 80 mmol of Zn and 30 mmol of Cu kg^{-1} wet wt. In undischarged nematocysts Cu and Zn are exclusively associated with the filament and therefore must also be bound or immobilised. The mature filament also had some 40 mmol kg^{-1} of Ca but it cannot be stated with confidence whether this is specifically associated with the filament material or is simply a reflection of some adherent capsule fluid.

Ionic Mechanism of Nematocyst Discharge. It is evidend from Table 1, that the elemental composition of the nematocyst changes drastically during the process of discharge induced by electrical stimulation in situ. The most dramatic change is in the loss of Ca from the capsule fluid and the gain of Na and Cl but not of SO_4. The values in Table 1 (cf. Lubbock et al. 1981) are averages from a number of nematocysts in different stages of discharge. The relationship between the loss of Ca and of dry mass from the capsule and the gain of Na and Cl is graphically depicted in Fig. 5, representing the changes in the acrorhagial nematocysts of *A elegantissima,* stimulated in stitu with the allogeneic column tissue. The data in Table 1 and in Fig. 5 show that these changes are the same in *R. rhodostoma* and *A. elegantissima* The final fluid in fully discharged nematocyst capsules, with filament everted, is essentially sea water. These changes in the ionic contents of the nematocyst in *R. rhodostoma* are pictorially depicted in Fig. 4 A–D, where the STEM image of a frozen dried section is correlated with the images from X-ray signals of Ca, K and Cl from the same field in

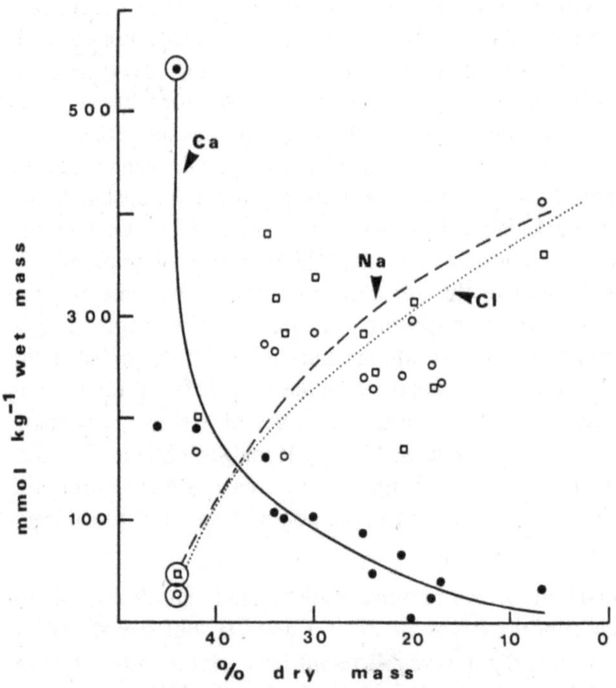

Fig. 5. *Anthopleura elegantissima.* Graphic representation of the changes in the major ions and dry mass during the discharge of nematocysts. (From Lubbock et al. 1981). • = Ca; □ = Na; ○ = Cl; ringed points represent undischarged state (n = 50)

the hydrated state of the section. X-ray quanta generated from Na were too weak to produce a corresponding image within a reasonable exposure time.

From these results Lubbock et al. (1981) proposed that the mechanism of nematocyst discharge in situ could be explained by the following chain of events. (1) The appropriate stimulus interacts with some cell surface receptors and induces electrical activity in the apical cell membrane (Lubbock and Shelton 1981). Incidentally, a change in the electrical and permeability properties of the nematocyte is reflected in a large influx of Na and Cl into the cell: a similar influx was also found in *A. elegantissima* (data not given here). (2) By some unknown mechanism, perhaps similar to the stimulus-response coupling in other secretory systems, the transduction of the signal leads to an exocytotic fusion of the apical cell membrane with the vacuole membrane enclosing the nematocyst, thus leading to the influx of sea water into the vacuole. (3) Simultaneously some other as yet unknown agent causes the release of Ca from the capsule fluid. This agent could conceivably be a change in pH caused by a Na^+/H^+ exchange or by proton pumping across the vacuole membrane now continous with the apical cell membrane. Such pH changes have been postulated for the egg activation during fertilization (Epel 1977, 1980) and for other exocytotic processes such as the discharge of chromaffin granules (Pollard et al. 1979). The change in the pH or some other factor shifts the binding constant of Ca to the peptide chains in the capsule fluid thus releasing the Ca^{2+}, disassociating the peptide chains and causing increase in the osmotic pressure of the capsule fluid. It should be noted that since the capsule wall is freely permeable to substances up to MW 700, ions could not be effective in increasing the osmotic pressure of the capsule contents: the impermeant peptides must play a primary part in this process. (4) An increase in the osmotic activity of the capsule contents leads to the influx of sea water, thus raising the intracapsular hydrostatic pressure and causing an explosive eversion of the filament as well as the injection of the fluid including the toxins through the tubular filament into the target. Picken and Skaer (1966) did mention that the freezing point of the capsular fluid from a nematocyst arrested during discharge was about $-1\,^{\circ}C$ lower than the lowest value recorded from the undischarged capsule.

An interesting observation (not reported by Lubbock et al. 1981) is that during eversion the filament loses most of its Zn but not its Cu component. At this stage of the nematocyst discharge some 20 mmol kg^{-1} wet wt. of Zn also appears in the capsular fluid as well as in the vacuole fluid. Zinc is not added to the artifical sea water. Thus it seems that in addition to the osmotic changes, the release of Zn from the filament might also be important in the discharge mechanism. Earlier Picken (1953, 1957) and more recently Carre (1980) have suggested that after the loss of operculum, possibly caused by the intracapsular hydrostatic pressure the eversion of the filament might be effected by the release of mechanical tension built into the folding of its components during the assembly. The allosteric contraints on the uneverted filaments might conceivably be provided by the salt-bridges mediated by Zn. Electrostatic interactions and salt-bridges do play a big role in the allosteric changes of the tertiary and quaternary structures of proteins both in monomeric systems and in supramolecular assembly and disassembly of biological structures (Perutz 1978).

Ionic Regulation of the Nematocyst Maturation. It is clear from Table 1 that the elemental composition of the pronematocyst in the nematoblast is very similar to that of surrounding cytoplasm and quite different from that of the undischarged nematocyst in the mature nematocyte. The mysterious events that lead to these drastic and complex ultrastructural changes have been a matter of continuos debate amongst biologists (Chapman and Tilney 1959; Slautterback and Fawcet 1959; Slautterback 1961; Chapman 1961; Skaer and Picken 1965; Skaer 1973; Germer and Hundgen 1980; Holstein 1980). We speculate the following sequence of events. (1) When all the macromolecular components of the nematocyst have been secreted, the vesicle membrane surrounding the pronematocyst is altered to incorporate Na^+/K^+ pumps. Activation of these pumps leads to the addition of Na^+ and removal of K^+. (2) A H^+-pump is incorporated into the vacuole membrane especially at the junction of capsule and the tube or over the entire tube. This could be provided by some very dense granules loaded with Cu and Zn and presumably like the secondary lysosomal "concretion" bodies invovled in heavy metal sequestration in a range of animals (Brown 1982). Cu and Zu-loaded granules are abundant in the cytoplasm of the nematoblast as well as in the other epidermal cells in both the tissues. It now seems established that the membranes of ER-Golgi derived vesicles do have H^+-pumps (Allen and Fok 1983). We have found that the Cu-Zn granules frequently adhered to the maturing nematocyst membrane at the front part. The image resolution in our microanalytical SEM does not resolve events such as exocytotic membrane fusion but it is likely to be the mechanism by which Cu and Zn together with the metalloproteins in the dense granules are added to the pronematocyst. (3) Addition of H, Na, Cu and Zn ions and metalloproteins promotes the formation of salt linkages and initiates the process of segregation and folding by the filament components leading to the invagination of the tube. (4) Since the Na ions appear to be sequestered by the filament material (see above) the removal of both K and Cl from the capsule is accompanied by the removal of water. Dehydration of the capsule contents would help in the condensation process of the proteins in the filament and the capsule wall as previously suggested by Picken (1957) and Skaer (1973). (5) Towards the end of maturation Ca ions are added, perhaps by the tubular differentiations of the vacuole membrane as discussed in Section 1.1 above (cf. Robson 1973).

The above proposals are speculative and the events need not necessarily occur in the suggested sequence. However, the suggested scheme would accommodate the data on elemental compostion of the component structures in the mature, undischarged nematocyst in *R. rhodostoma.* It is also consistent with what is known about the self-assembly of proteins in general (Perutz 1978). For example, the protonation of protein disks triggers their assembly into the helical rods of tobacco mosaic virus. Weakly ionised carboxyl groups in peptides (pk < 3) can preferentially bind Na^+ while strongly ionised –ve charges as in sulphated glycoproteins (collagen of the capsule wall) preferentially bind K (Scott 1978). Incorporation of Cu and Zn ions has been shown to have a specific role in stabilizating the quartenary sturcture of many enzymes (Getzoff et al. 1983).

3 Conclusions

The electron microprobe X-ray analysis of nematocysts in situ carried out an 1–2 μm thick cryosections of fresh-frozen mesenterial filaments of *R. rhodostoma* and acrorhagi of *A. elegantissima* has revealed a key role of ionic changes during the differentiation, maturation and discharge of nematocysts. The results show unambiguously the role of some 600 mmol kg^{-1} wet wt. of Ca in the osmotic discharge of capsular contents. The analysis has allowed for the first time an insight into how the structurally most complex secretion product of the ER-Golgi system might be self-assembled by condensation, protonation and salt-linkages provided by Na^+, Cu^{2+} and Zn^{2+} ions at concentrations of 50–100 mmol kg^{-1} wet wt. The results also imply that a complete mechanism of nematocyst discharge may not be only osmotic but may also involve a kinetic change in the filament structure due to the release of allosteric restraints by the removal of Zn^{2+}. Clearly a great deal more remains to be learnt about this fascinating problem. It will be appropriate to recall a comment made by Don Fawcett some 25 years ago in another symposium (Slautterback 1961), where he suggested that the mechanism of tube inversion during nematocyst maturation is the reverse of the firing mechanism. The converse would also hold. Our microprobe results now lend credence to Fawcett's remarks.

References

Alberts B, Bray D, Lewis J, Martin R, Roberts K, Watson JD (1983) Molecular biology of the cell. Garland, New York

Allen RD. Fok AK (1983) Nonlysosomal viscles (acidosomes) are involved in phagosome acidification in *Paramecium.* J Cell Biol 97: 566–570

Beckett A, Porter R. Read ND (1982) Low temperature scanning electron microscopy of fungal material. J Microsc (Oxf) 125: 193–201

Blanquet R (1968) Properties and composition of the nematocyst toxin of the sea anemone *Aiptasia pallida.* Comp Biochem Physiol 25: 893–902

Blanquet R (1970) Ionic effects on discharge of the isolated and in situ nematocysts of the sea anemone *Aiptasia pallida*: a possible role of calcium. Comp Biochem Physiol 35: 451–461

Brown BE (1982) The form and function of metal-containing 'granules' in invertebrate tissues. Biol Rev (Camb Philos Soc) 57: 621–669

Carre D (1980) Hypothese sur le mecanisme de l'evagination due filament urticant des cnidocytes. Eur J Cell Biol 20: 265–271

Chun C (1881) Die Natur und Wirkungsweise der Nesselzellen bei Coelenteraten. Zool Anz 4: 646–650

Chapman GB (1961) The fine structure of the stenoteles of Hydra. In: Lenhoff HM, Loomis WF (eds) The biology of Hyda and some other coelenterates. University of Miami Press, Coral Gables, P 131

Chapman GB, Tilney LG (1959) Cytological studies of the nematocysts of Hydra. J Biophys Biochem Cytol 5: 68–84

Conklin EJ, Mariscal RN (1976) Increase in nematocyst and spirocyst discharge in a sea anemone in response to mechanical stimulation. In: Mackie GO (ed) Coelenterate ecology and behavior. Plenum, New York, p 549

Epel D (1977) The program of fertilization. Sci Am 237 (II): 128–138

Epel D (1980) Fertilization. Endeavour (New Series) 4: 26–31

Germer I, Hündgen M (1980) The fine structure of developing and mature cnidoblasts of Eirene bircidula. In: Tardent P and Tardent R (eds) Developmental and cellular biology of coelenterata. Elsevier/North Holland, Amsterdam, p 325

Getzoff ED, Tainer JA, Weiner PK, Kollman PA, Richardson JS. Richardson DC (1983) Electrostatic recognition between superoxide and copper, zinc superoxide dismutase. Nature (Lond 306: 287–290

Gupta BL (1976) water movement in cells and tissues. In: Spencer-Davies P (ed) Perspectives in experimental biology, vol I. Pergamon, Oxford, p 25

Gupta BL, Hall TA (1978) Electron microprobe X-ray analysis of calcium. Ann NY Acad Sci 307: 28–51

Gupta BL, Hall TA (1981) The X-ray microanalysis of frozen-hydrated sections in scanning electron microscopy: an evaluation. Tissue & Cell 13: 623–634

Gupta BL, Hall TA (1982) Electron probe X-ray microanalysis. In: Baker PF (ed) Technique in cellular physiology –part II. Elsevier/North-Holland, Amsterdam County Clare, p 128/1

Hall TA, Gupta BL (1983) The localization and assay of chemical elements by microprobe methods. Q Rev Biophys 16: 279–339

Holstein T (1980) An ultrastructural and kinematographic study of the nematocyst morphogenesis in selected coelenterate species. In: Tardent P and Tardent R (eds) Developmental and cellular biology of coelenterata. Elsevier/North-Holland, Amsterdam, p 313

Jaffe LF (1983) Sources of calcium in egg activation: a review and hypothesis. Dev Biol 99: 265–276

Lane CE (1968) Cemical aspects of ecology, pharmacology and toxicology. Chem. Zool 2: 263–284

Lenhoff HM, Kline ES, Hurley R (1957) A hydroxyproline-rich, intracellular, collagen-like protein of *Hydra* nematocysts. Biochim Biophys Acta 26: 204–205

Lubbock R (1980) Clone-specific anemone. Proc Natl Acad Sci USA 77: 6667–6669

Lubbock R, Amos WB (1981) Removal of bound calcium from nematocyst contents causes discharge. Nature (Lond) 290: 500–501

Lubbock R, Shelton GAB (1981) Electrical activity following cellular recognition of self and non-self in a sea-anemone. Nature (Lond) 289: 59–60

Lubbock R, Gupta BL, Hall TA (1981) Novel role of calcium in exocytosis: mechanism of nematocyst discharge as shown by X-ray microanalysis. Proc Natl Acad Sci USA 78: 3624–3628

Mariscal RN (1974) Nematocysts. In: Muscatine L and Lenhoff HM (eds) Coelenterate biology: reviews and new perspecitves. Academic, New York, p 129

Mariscal RN (1980) The elemental composition of nematocysts as determined by X-ray microanalysis. In: Tardent P and Tardent R (eds) Developmental and cellular biology of coelenterates. Elsevier/North-Holland, Amsterdam, p 337

Mariscal RN, Conklin EJ, Bigger CH (1978) The putative sensory receptors associated with the cnidae of cnidarians. SEM/1978/II: 959–966

Mariscal RN, Mclean RB (1976) The form and function of cnidarian spirocysts. 2. Ultrastructure of the capsule tip and wall and mechanism of discharge. Cell Tissue Res 169: 313–321

Perutz MF (1978) Electrostatic effects in proteins. Science (Wash DC) 201: 1187–1191

Picken LER (1953) A note on the nematocysts of *Corynactis viridis.* Q J Microsc Sci 94: 203–227

Picken LER (1957) Stinging capsules and designing nature. New Biol 22: 56–71

Picken LER, Skaer RJ (1966) A review of researches on nematocysts. Symp Zool Soc Lond 16: 19–50

Pollard HB, Christopher JP, Creutz CE, Zinder O (1979) The chromaffin granules and possible mechanisms of exocytosis. Int Rev Cytol 58: 159–197

Rasmussen H (1981) Calcium and cAMP as synarchic messengers. Wiley Interscience, New York

Robson EA (1953) Nematocysts of *Corynactis*: the activity of the filament during discharge. Q J Microsc Sci 94: 229–235

Robson EA (1973) The discharge of nematocysts in relation to properties of the capsule. Publ Seto Mar Biol Lab 20: 653–665

Ross KFA (1967) Phase contrast and interference microscopy for cell biologists. Edward Arnold, London

Rubin RP (1982) Calcium and cellular secretion. Plenum, New York

Salleo A (1983) Present symposium

Scott JE (1978) Some comments on specific ionic interactions. In: Everett DH and Vincent B (eds) Ions in macromolecule and biological systems. Scientechnica, Bristol, p 132

Skaer R (1973) The secretion and development of nematocysts in a siphonophore. J Cell Sci 13: 371–393

Skaer PJ, Picken LER (1965) The structure of the nematocyst thread and the geometry of discharge in *Corynactis viridis* (Allman). Philos Trans R Soc Lond B Biol Sci 250: 131–164

Slautterback DL (1961) Discussion. In: Lenhoff HM and Loomis WF (eds) University of Miami Press, Coral Gables, p 127

Slautterback DB (1967) The cinoblast-muscloepithelial complex in the tentacles of Hydra. Z Zellforsch Mikrosk Anat 79: 296–318

Slautterback DL, Fawcett DW (1959) Development of the cnidoblasts of *Hydra*. An electron microscope study of cell differentation. J Biophys Biochem Cytol 5: 441–452

Tardent P, Tardent R (1980) Developmental and cellular biology of coelenterates. Elsevier/ North-Holland, Amsterdam

Weill R (1930) Essai d'une classification des nematocystes des cnidares. Bull Biol Fr Belg 64: 141–153

Williams RTP (1977) Calcium chemistry and its relation to protein binding. In: Wassermann RH Carradino RA, Carafoli E, Ketsinger RH, Maclennan RH and Siegel FL (eds) Calcium-binding proteins and calcium function. North-Holland, New York, p 3

Wolf AV (1966) Aqueous solutions and body fluids: their concentrative properties and conversion tables. Harper Row, New York

Yanagita TH (1943) Discharge of nematocysts. J Fac Sci Univ Tokyo. Sect V Zool 6: 97–108

[illegible] (1967) [illegible] and interference [illegible] in [illegible] cell biology [illegible]
[illegible] London
[illegible] of [illegible] Plenum, New York
[illegible] symposium
[illegible] (19[illegible]) Some [illegible] to [illegible] in [illegible]
[illegible] anatomy [illegible] and biology [illegible]
[illegible] development of [illegible]
[illegible]

Part B
Responses of Marine Animals to Pollutants

Respiratory Adaptations of Marine Fish in Relation to Their Mode of Life Under Different Environmental Conditions

G.M. HUGHES[1]

1 Introduction

The uptake of pollutants into the fish body occurs by three main routes – the skin, food and gills (Brown 1976). The relative importance of these three sites varies both in individual species and according to the nature of the pollutant. The present discussion is concerned entirely with uptake at the gills.

The respiration of fish is closely related to the physico-chemical properties of their respiratory medium (Hughes 1961; Dejours 1976), which is normally water. In particular its low oxygen content relative to a similar volume of air increases the volume of water which must be pumped across the gas exchange surfaces in unit time if sufficient oxygen is to be extracted to meet the metabolic requirements of the animal (Table 1). As at 15°C water is about 800 times more dense and 60 times more viscous than air, this must entail greater energy expenditure by water-breathing animals. Although most respiratory adaptations of fish have been considered in the context of oxygen supply, nevertheless, the high solubility of CO_2 in water (30 times

Table 1. Goldfish. Relationship between temperature, O_2 content of water and ventilation volume necessary to satisfy resting O_2 uptake ($\dot{V}_{O_2}$) and consequent increase in ventilatory requirement. (After Hughes 1963, based on data of Fry and Hart 1948)

Temp	$\dot{V}_{O_2}$ Resting	$\dot{V}_{O_2}$ Active	O_2 content of water	$\dot{V}_G$ required (U = 75%)	Ventilatory requirement $\dot{V}_G/\dot{V}_{O_2}$
°C	(ml kg^{-1} h^{-1})		(ml l^{-1})	(l kg^{-1} h^{-1})	(ml $H_2O \cdot l^{-1} O_2$)
5	8	30	9.0	1.3	0.16
15	50	110	7.0	9.0	0.18
25	140	255	5.0	32	0.23
35	225	285	5.0	60	027

1 Research Unit for Comparative Animal Respiration, University of Bristol, Woodland Road, Bristol B58 IVG, England

Toxins, Drugs, and Pollutants in Marine Animals
(ed. by Bolis et al.)

that of O_2) renders it a more favourable medium for release of CO_2 which has also been an important factor during the evolution of vertebrates. For example, studies of the transition from aquatic to aerial breathing in bimodal breathers have revealed many organisms which obtain most oxygen from the air but liberate carbon dioxide into the water (Rahn and Howell 1976; Singh 1976). Possible advantages of such a strategy were envisaged many years ago (Hughes 1966b; Rahn 1966).

In this context it is clear that the aquatic environment is by no means uniform. The marine environment is generally regarded as more constant than freshwaters both with respect to gas and ionic composition and also the range of temperatures to which animals are subjected in a particular region. Nevertheless, there are parts of the marine environment where changes in such physico-chemical features may be quite large, notably in transitional zones between aquatic and terrestrial and between sea and freshwaters. There are, however, also regions of the ocean where the oxygen content is extremely low (Macdonald 1972; Douglas, Friedl, Pickwell 1976) and others in which it is supersaturated such as antarctic oceans (Holeton 1980), splash zones and rockpools in which the P_{O_2} may exceed 200 mmHg at certain times of the day. Some of the benthic habitats occupied by fish also have very low oxygen content and have given rise to numerous adaptations including the use of anaerobic mechanisms (Fig. 1).

In relation to this range of environments, adaptations are present at morphological, physiological and biochemical levels and at all parts of the respiratory chain (Hughes 1964, 1973). Because of these differences between fish, it is important to appreciate that different species might be affected to different degrees by given pollutants not only because of the specific action of the pollutant, but also because a given pollutant concentration could have more serious consequences in some environ-

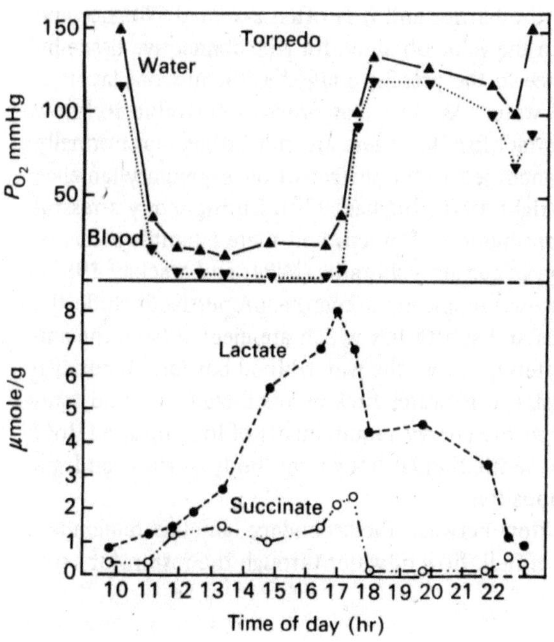

Fig. 1. *Torpedo marmorata.* Changes in blood onstituents during a deep hypoxia (inspired P_{O_2} = 0–5 mm Hg) and their recovery following return to normoxic conditions. Note the increase in lactate concentration but also the gradual build-up of succinate during hypoxia. Normal succinate levels are recovered very rapidly whereas lactate concentrations do not return until 18 h later. (After Hughes and Johnston 1978)

ments than in others. Furthermore, because the normal environmental range to which some fish are subjected may be greater than others, so these species may be more capable of resisting adverse respiratory conditions arising from pollutant action. In this context it is of interest to consider different parts of the respiratory systems of fish and to emphasise some of the ways in which such adaptations interact with the severity of any pollutant.

2 Gills and their Ventilation

Fish gills show the finest adaptations among aquatic organisms for the extraction of the relatively small quantities of oxygen. To do this pumping, mechanisms have evolved which maintain a more or less continuous flow of water across the finely divided respiratory surfaces. The secondary lamellae provide the exchange surface between water and blood and some of the water/blood distances may be less than 1μm in very active fish (Hughes 1970). Similarly, such species have a large gill area and these two features combine to give a large O_2-diffusing capacity.

$$\begin{array}{c}\text{Diffusing capacity}\\ \text{(physiological)}\end{array} \quad \frac{\dot{V}_{O_2}}{\Delta P_{O_2}} = \frac{K.A}{t} \quad \begin{array}{c}\text{Diffusing capacity}\\ \text{(morphometric)}\end{array} \qquad (1)$$

Thus from a physiological point of view, diffusing capacity is the volume of gas which is transferred in unit time ($\dot{V}_{O_2}$) for a mean difference of unit oxygen tension, and this may be equated to the morphometric diffusing capacity which takes into account the area of the exchange surface (A) and the effective barrier thickness (t) together with a suitable coefficient which is the Krogh permeation coefficient (K). The barrier itself comprises many different layers some of which are illustrated in Fig. 2 and 5. Unfortunately very little information is available regarding the permeation coefficient of these different parts of the barrier and it is often assumed (Hughes and Perry 1976) that they are all equal to the value obtained for frog connective tissue by Krogh (1919). Of particular relevance in the present context is the mucous layer separating the tissue barrier from the water. As yet it has proved impossible to know the condition of this barrier in the intact fish, but there are indications that normally there is a very thin film of mucus anchored to the surface of the secondary lamellae by the microridges (Hughes and Wright 1970; Hughes1979). During many stressful conditions fish show an increased production of mucus and there is ample evidence for this takingplace during the action of certain pollutants (Miller and Mackay 1982). Mucus has a complex and varied structure and its numerous properties include the ability to "mop-up" toxic organisms and substances which are ejected from the animal and thereby prevented from entering across the water/blood barrier. Mucus also has an important effect on the resistance to water flow across the gill and body surfaces and may reduce drag and hence the energy requirements of locomotion (Hoyt 1975; Daniel 1981) and probably ventilation although the body surface and gill mucus appear to have different compositons.

Details of the patterns of water flow between the secondary lamellae remain largely unknown, the general pattern of bulk flow of water through the respiratory sys-

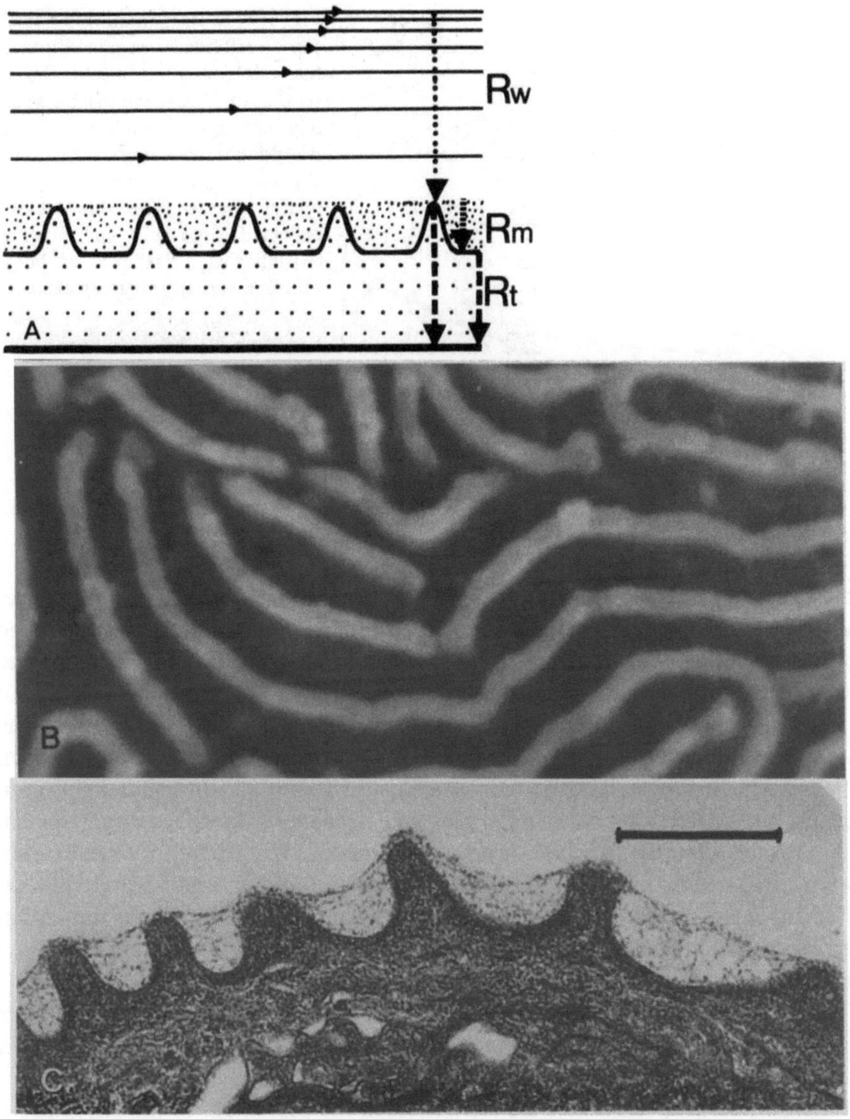

Fig. 2. A Diagram to illustrate the probable nature of the gill surface with a series of microridges between which is present a film of mucus. **B** Scanning electron micrograph of the surface of a secondary lamella of rainbow trout (× 24.000). **C** High power transmission electron micrograph through the surface of the epithelium of the secondary lamella of a flatfish (Solea Variegata); scale = 1 μm. (**A** and **B** after Hughes 1979; **C** after Hughes and Wright 1970)

tem has been established by recording the pressure changes during a ventilatory cycle. Such studies (Hughes and Shelton 1962; Hughes 1960) showed that this is achieved by a double pumping mechanism involving rhythmic changes in volume of the cavities on both sides of the gill resistance (Fig. 3). Thus water flow is maintained partly by the greater pressure produced by the buccal pumping mechanism together with the suction action of the opercula and branchiostegal apparatus. The relative balance

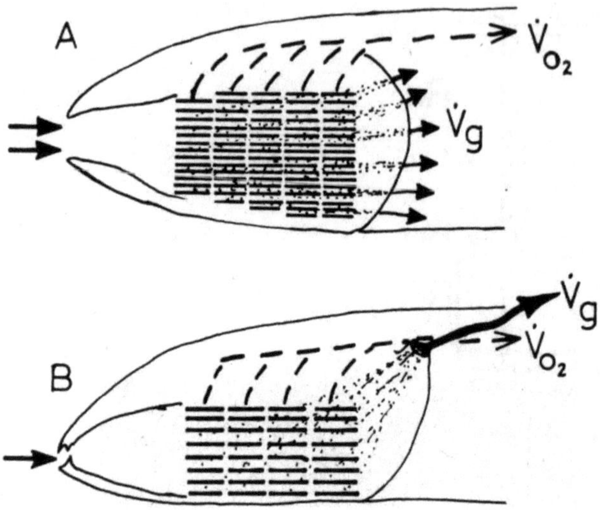

Fig. 3 A, B. Diagrammatic comparison of differences in dimensions of the gills and their mechanism of ventilation in **A** active species which rely on ramjet ventilation and **B** sluggish bottom-living species which depend mainly on opercular suction pumps. Dashed line shows blood collected from gills and provides oxygen ($\dot{V}_{O_2}$) for rese of body. Full arrows are ventilatory current

of these two pumps varies in different environments (Table 2) – in general species living on the sea bottom have their opercular suction mechanisms better developed than in midwater species in which the two pumps are more or less balanced (Hughes 1960). Most pelagic species depend far more on the buccal pumping mechanism and this reaches its ultimate development in fish which depend almost entirely on the greater pressure developed in the buccal cavity as a result of their forward propulsion (Roberts 1975; Brown and Muir 1970; Graham and Laurs 1982). In such ramjet ventilation the propulsive musculature (Rayner and Keenan 1967; Roberts and Graham 1979) provides the energy which forces water through the respiratory cavities and between the secondary lamellae. The flow is almost continuous and reduces the extent of any dead spaces within the ventilatory system. Measurements of ventilation have been carried out in a number of species and values for minute volume have been determined both by direct and indirect means (Hughes and Knights 1968; Dejours 1973; Johansen 1982). In all fish the volume of water to be pumped across the gills to obtain unit volume of oxygen is far greater than that in aerial organisms. This is sometimes called the ventilatory requirement ($\dot{V}_G/\dot{V}_{O_2}$) (Table 1).

Among different species there are differences in the percentage of the oxygen contained in the ventilated water which is removed by the gill system and clearly those in which this percentage utilisation (U) is greater will have a lower ventilatory requirement. In most fishes the percentage utilisation decreases (Fig. 4) with increase in ventilation volume (Hughes and Umezawa 1968; Shelton 1970) largely due to the effect of the increased water flow on the morphological distribution of the gill system. Consequently there is an increase in water shunted between the gill filaments of adjacent arches and the more rapid flow of water between secondary lamellae increases the physiological dead space (Hughes 1966a; Randall 1970). But special modifications which consolidate the gills of some tunas reduce this effect and utilisations of 50% may be present (Stevens 1972). In general, however, the more rapidly swimming species have a higher ventilation and lower utilisation than the more

Table 2. Relationship between mode of life, structure of the gills, and ventilation mechanism of marine fishes

Mode of life	Respiratory function	Habitat		Ventilation mechanism		Example
Active	Large gill area Secondary lamellae closely spaced High ventilation volume	Midwater	Swimming	Continuous	Ramjet	Tuna
				Intermittent	Buccal and opercular pumps balanced	Wrasse
			Hovering			*Balistes*
Inactive	Small gill area Low secondary lamellar frequency Low ventilation volume	Seabottom			Opercular suction pump dominant	Dragonet
			Buried		Low ventilatory frequency	Flatfish

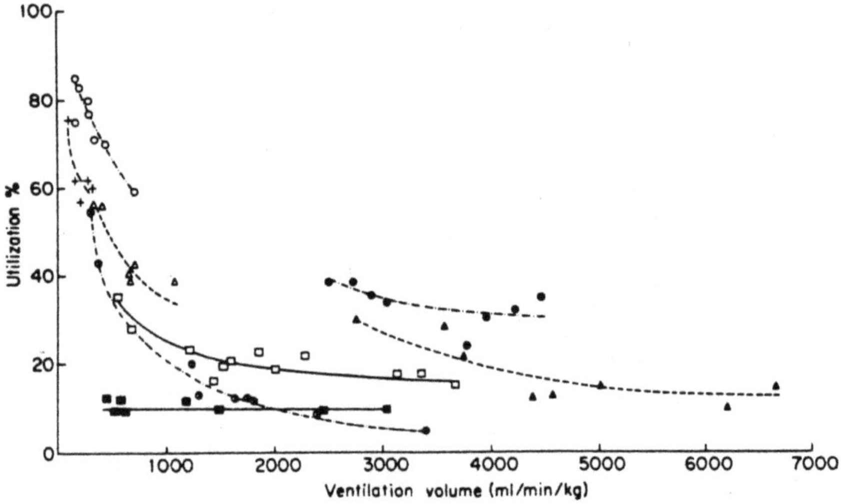

Fig. 4. Relationship between utilization and ventilation volume in fish under different environmental conditions. The conditions which led to increased ventilation are specified below. Carp (Saunders 1962) 20°C. (○) hypoxia; (●) exercise, and (◑) carbon dioxide excess and hypoxia. Sucker (Saunders 1962) 20°C; (△) hypoxia and (▲) exercise. Trout (Holeton and Randall 1967) 15°C, (□) hypoxia; and (Stevens and Randall 1967) 5°C, (■) exercise. Dogfish (Baumgarten-Schumann and Piiper 1968) 16°C, (+) recovery from activity (from Shelton 1970)

benthic species where water flow through the gills is slower and a larger utilisation is obtained. Very high values (>90%) of utilisation and low ventilation volumes are present in midwater "hoverers" such as *Balistes* (Hughes 1967). Equations relating oxygen uptake ($\dot{V}_{O_2}$), ventilation volume ($\dot{V}_G$) and utilisation (U) are as follows:

$$\dot{V}_{O_2} = \dot{V}_G \; (P_{insp} - P_{exp}) \, \alpha_{O_2} \qquad (2)$$

$$U\% = \frac{(P_{insp} - P_{exp})}{P_{insp}} \times 100 \qquad (3)$$

Pinsp and Pexp are the partial pressures of oxygen in the inspired and expired water, α_{O_2} is the solubility or absorption coefficient of oxygen in water. The product $\dot{V}_G \cdot \alpha_{O_2}$ is the capacity rate for oxygen and defines the quantity of oxygen passing the gills in unit time and is generally higher for more active fish.

Against this general background of fish respiratory physiology, it is of interest to consider the situation when the water contains other substances including pollutants and the extent to which they may be considered in a way comparable to that in which oxygen has been studied. Thus, it is clear that the larger the ventilation volume the greater the quantity of pollutant passing between secondary lamellae. Much will depend upon the particular nature of the pollutant as to the seriousness of this effect. If the pollutant is completely in solution, then presumably it would behave in a way similar to the oxygen molecules. In some cases this would result in a larger total quantity of pollutant passing between the lamellae but perhaps the utilisation would be lower as a consequence of a larger proportion of the pollutant passing along what corresponds to the dead space for oxygen. On the other hand, suspended and particulate material might have a far more serious effect if the dimensions were such that the increased ventilation volume would bring a larger number of particles into direct contact with the gill surfaces especially as the lamellar sieve is finer. In general, then, one would suppose that active fish which ventilate a far larger volume and have a higher oxygen uptake would also be subjected to greater exposure to pollutants. Benthic fish generally have lower ventilation volumes, coarser gill sieves and higher utilisations. Consequently, with respect to a dissolved pollutant, they probably absorb far more per unit ventilation volume than would more active organisms, but the total quantity of pollutant to which they would be subjected would be smaller.

Considerations of this kind also draw attention to the importance of the pollutant equivalent to the solubility coefficient for oxygen (α_{O_2}) which is present in Eq. (2) above. Thus for some pollutants transfer across the water/blood barrier would be governed by the same basic relationship as for oxygen and partial pressure (or concentration) differences would be important, but also the pollutant's solubility in different parts of the barrier which would determine the equivalent of the Krogh permeation coefficient ($K=\alpha_{pol} \cdot D$, where D is the diffusion coefficient). From this point of view it would be valuable to have information regarding the solubility of different pollutant molecules in mucus and gill tissue as well as in sea water and blood. For smaller molecules D would also be greater, but for many pollutants, including divalent metal ions, the great hydration would produce a larger particle which would have D,

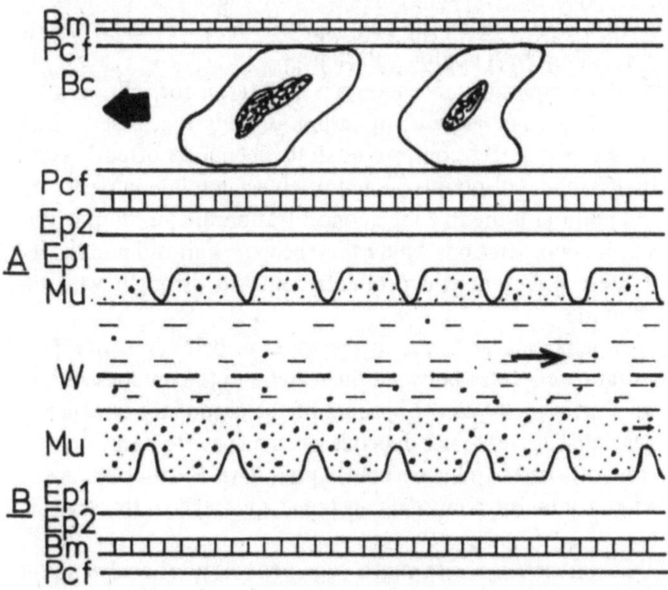

Fig. 5. Diagrams to illustrate possible differences in thickness of portions of the water/blood barrier (A) under normal conditions and (B) following exposure to pollutants (larger dots). In particular the thickness of the mucous film (*Mu*) increases and diffusion of pollutant between the water and blood is delayed. Much of the pollutant becomes lost as a larger quantity of mucus is swept away by the water current. *Bc* blood channel, *W* water, *Bm* basement membrane, *Ep* epithelial layers, *Pcf* pillar cell flange

and consequently K, values which would be less than for oxygen. As indicated above, however, a pollutant's action may be entirely or partially due to insoluble components, and these would be governed by different relationships.

The two diagrams in Fig. 5 indicate possible conditions at the water/blood barrier before and following ventilation of the gills with a pollutant. In the first instance, the thickness of the mucous layer is confined to the troughs between the microridges and the main barrier to oxygen transfer is provided by the tissue layers. Within the capillaries a red blood cell is shown being deformed during its passage, and this is a further variable, which can be affected by environmental conditions, but so far has been given little consideration (Hughes et al. 1982). Following exposure to a pollutant the mucous film will increase considerably both in volume and thickness and hence reduce dimensions of the interlamellar spaces which will increase the resistance to water flow. Although the resistance to oxygen transfer will also be increased, the relationship between this and the resistance to pollutant transfer will depend very much on the permeability properties of the mucous layer relative to oxygen. An ideal mucus would have very high permeability to oxygen but low permeability to pollutant and, as discussed above, this would probably be true for many heavy metals, the divalent ions of which are usually in the hydrated form. Because of a low permeation coefficient in the mucus there would be a tendeny for the pollutant to accumulate to a relatively high concentration in this layer and the continuous washing-off of excess mucus would provide a protective mechanism, reducing the amount of pollutant which penetrated to the epithelial surface and thence to the blood. Such a strategy of accumulation, in this case in a layer outside the body, is well known within living organisms as a mechanism for dealing with noxious substances.

Again, a comparison with the respiratory mechanism is of interest for following the uptake of oxygen a more or less equal amount of carbon dioxide is released which diffuses very rapidly into the water. By contrast noxious chemicals often become metabolised within the body, rendered non-toxic and often accumulate in particular tissues. However, when the pollutants are in a high concentration the build-up within the living organism may reach concentrations where they become harmful and indeed lethal. Low solubility of these pollutant end products contrasts strongly with the high solubility of carbon dioxide.

Narrowing of the inter-lamellar spaces increases the water velocity and would improve the transfer of pollutant molecules between the water and the mucous surface just as oxygen transfer is improved by closer spacing of the secondary lamellae in more active fish. It is clear that a great deal depends on the specific nature of the interaction between the mucous layer and a particular pollutant. Further studies in this field are therefore necessary before a more quantitative approach to this transfer can be developed.

The main purpose of this contribution is to suggest the possibility of analysing the gas exchange pathway for oxygen in terms of the physiochemical properties of different pollutant molecules. Some of the basic equations [e.g., Eq. (2)] at present used in fish respiratory physiology were obtained by analogy with equations used by engineers for heat exchangers (Hughes and Shelton 1962; Hughes 1964). This was appropriate as heat exchangers can have counter-current or parallel current arrangements. It is now suggested that the same equations, as modified by respiratory physiologists, are applicable to the uptake of pollutants at the gills, for example:

$$\dot{V}_{pol} = \dot{V}_G \alpha_{pol} (P_{pol,\, insp} - P_{pol,\, exp})$$

Such a discussion is very speculative at this stage, but is given here in the hope that it might stimulate some more quantitative approaches to this problem.

3 The Gas Transport System

Following the transfer of oxygen across the water/blood barrier, the oxygen molecules come into a chemical combination with haemoglobin in the red blood corpuscles. The rate of reaction berween the oxygen and haemoglobin is a part of the resistance to gas transfer and correspondingly some pollutant molecules influence this part of the respiratory chain. Others may simply stay in solution in the plasma and be transported to other parts of the fish where the most toxic action would occur. Again, by analogy with oxygen transport one can suppose that blood would have certain dissociation characteristics so that pollutants would be given up in regions where they were in lowest concentration, assuming a purely physical transfer mechanism. Equally, however, they might tend to aggregate by direct combination with certain parts of the microcirculation. Thus the physicochemical characteristics of any part of the vascular system could be modified and the red blood cell membranes would seem to be the most exposed areas.

Corresponding to the release of carbon dioxide, one can think of detoxification mechanisms whereby noxious chemicals are released into the blood and pass to the exchange surfaces, where they are lost to the environment. But in other cases there would be an accumulation of such chemicals or they might be released into the kidney ducts. Different fish have adopted different strategies, and it seems difficult to make any generalisation relating to the particular environment and habits of different species.

4 Direct Action of Pollutants on the Gas Exchange System

In addition to the relationships discussed above, there have been many studies which indicate that some pollutants can directly affect the gas exchange mechanisms of fish and their toxic action is a secondary product of hypoxia rather than a primary effect of their presence in the blood. Notable examples are some heavy metals which damage the surface of secondary lamellae (Skidmore and Tovell 1972) and reduce their diffusing capacity as a result of an increase in diffusion distance and a reduction in effective area (Hughes 1976a). From morphometric studies (Hughes and Perry 1976), a convenient way of representing this situation was by introducing the term "relative diffusing capacity", i.e., the relationship between the diffusing capacity of the experimental or treated fish compared with that of controls. The histograms in Fig. 6 illustrate the way in which following exposure to nickel the diffusing capacity is

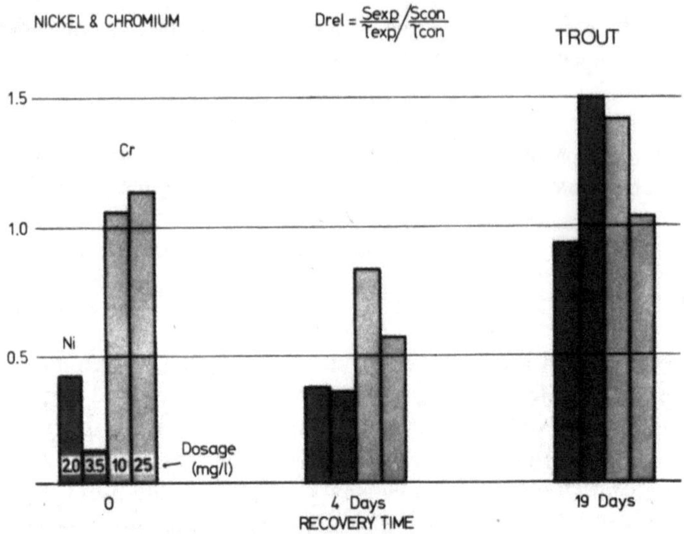

Fig. 6. Histograms to show changes in relative diffusing capacity (D_{rel}) of rainbow trout following treatment for 5000 min with the concentrations of nickel and chromium indicated in the columns. Recovery following marked reduction in diffusing capacity after nickel contrasts with the relatively slight changes following chromium pollution. (After Hughes 1976 a)

about one half that of normal fish and hence D_{rel} = c.0.5 Following recovery over 19 days this returns to normal levels (i.e., 1.0). The effect on D_{rel} and the rate of recovery is related to the sublethal concentration of the nickel. For other heavy metals, e.g., chromium, changes of this kind were not observed; in the case of a very low concentration of cadmium D_{rel} was found to increase, suggesting an improvement in the gas exchange conditions of the secondary lamellae (Hughes, Perry and Brown 1979). Other pollutants have been shown to have direct effects on the haemoglobin gas transport system (Larsson 1973). In all these cases the considerations discussed elsewhere in this paper are also applicable.

5 Relation to Body Size

There is now much data on the relationship between oxygen consumption and body mass (Hemmingsen 1960; Schmidt-Nielsen 1970; Heusner 1983) and between gill area and body mass (Hughes 1984a, b) but few studies have shown effects of body size on the effects of pollutants. The well-known relationship between oxygen consumption and body mass, $\dot{V}_{O_2} = aW^b$, has been derived from both intra and interspecific studies of many fishes (Winberg 1956). Although in some cases the oxygen uptake is almost directly proportional (b=1.0) to body mass, in most cases the exponent is in the range 0.5 – 0.85. Thus smaller fish have a higher oxygen consumption per unit body weight than do larger specimens of the same species. This almost certainly involves a relatively greater increase in ventilation volume of smaller individuals. With the larger volume of water passing across the gill epithelium it is clear that smaller specimens will be subjected to a greater quantity of pollutant per unit time ($\dot{V}_G \alpha_{pol}$) than would be a larger fish. It would seem, therefore, that the total quantiy of pollutant which might be absorbed, assuming a constant permeability of the gill surface, would be greater for fishes with greater oxygen uptake per unit body mass.

In general there is also a similar relationship between gill surface area and body mass, although the exponents for the gill area and oxygen uptake relationships are not always the same (Hughes 1972). There is some evidence (Pauly 1981) that species with more rapid growth rates have higher exponents in the gill area relationship. For a given species the gill area per unit mass is greater among smaller specimens and once again this would tend to indicate a greater capability for pollutant uptake by smaller specimens. For a number of fish species it has been shown that the ratio of oxygen uptake to gill area ($\dot{V}_{O_2}/A$) is more or less constant (Hughes 1977) indicating that the mean difference of P_{O_2} across the water/blood barrier is also fairly constant among different species. Because of the constant $\dot{V}_{O_2}/A$ ratio it might be supposed that different species would be equally sensitive to water-borne pollutants. However, this is not true because the absolute values for oxygen uptake are far greater in more active than less active species, either in midwater or on the seabottom. In fact from the assumed constant ratio it has been possible to predict that fish such as *Latimeria* with low gill are would have a low oxygen consumption (Hughes 1976b).

These thereotical considerations do not seem to have been tested a great deal although there is some limited evidence for them. Among mammals a relationship has been established, particularly with respect to the action of anaesthetics. It is now appreciated that the dosage of drug or anaesthetic to be used should not be expressed

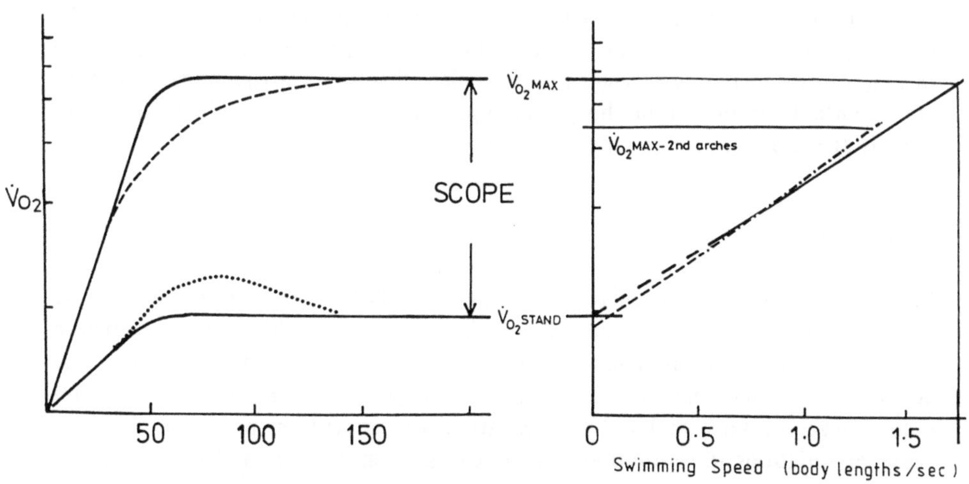

Fig. 7. Diagram illustrating the relationship between scope for activity in a rainbow trout whose oxygen consumption was determined over a range of swimming speeds. ***Dash-doted lines*** show similar plot for fish from which second gill arches had been removed. In the latter case the values for standard oxygen consumption (extrapolation to zero speed) are insignificantly different to those of control fish. Maximum $\dot{V}_{O_2}$ is less in fish with gill damage, however, which consegently have a lower aerobic scope. (After Hughes 1984a, Duthie and Hughes 1982)

per unit body mass but rather in relation to $W^{0.75}$ (Schmidt-Nielsen 1970). In preliminary studies with rainbow trout, I have found that smaller specimens certainly are anaesthetised more deeply when subjected to a concentration of MS222 which in a larger specimen (400 g) would only result in a restriction of locomotory movements, while ventilation continued. In some recent studies with zinc (Hughes and Tort 1983) when exposing fish to a lethal concentrations (20 ppm) but for a short period (2–2.5 h) indications were found that survival following return to clean water was greater for the larger individuals, thus confirming the view that the smaller trout had been subjected to a greater exposure to the zinc at their exchange surfaces. More detailed observations of this kind are clearly necessary before any quantitative assessment can be made.

From the known studies on oxygen consumption and gill area in relation to body mass, it is clear that different species will behave differently in this respect just as the relationship between body mass and scope for aerobic activity varies (Hughes 1984a). There seems to be some relationship between aerobic scope and the degree of damage to the gills of fish. Thus removal of the gill filaments from a single gill arch on each side of a rainbow trout (Duthie and Hughes 1982) reduces the surface area by approximately 30% and such specimens, although having approximately the same standard oxygen consumption, show a reduction in $\dot{V}_{O_2 max}$. (Fig. 7). The effect of damage to the gills produced by pollutant action is not so clear-cut for although some specimens have their swimming performance severely affected in others this was not so, although the duration and condition of exposure to zinc were identical. Variability between individual specimens is always apparent in studies involving survival under lethal conditions and seems to be related to the particular type of response that the specimens show during the stress. Rainbow trout subjected to a lowering of the environmental oxygen show a complete range of response patterns. Some specimens pursue rapid movements chiefly towards the water surface even when the P_{O_2} is only lowered to 90 mm Hg, whereas other specimens show very little change in locomotory behaviour even when the P_{O_2} is less than 40 mm Hg. Presumably there are important differences in the physiological status of the individuals, but as yet there is little evidence concerning their nature.

6 Conclusions

From this survey an approach to the analysis of pollutant action in marine fishes is suggested which takes into account the marked differences in life habits and environments occupied by different species. The ways in which their respiratory systems are adapted to different habitats entails different degrees of exposure to a pollutant, even though the same ambient concentration was present. In particular this is because of variations in ventilation volume, oxygen consumption and gill morphometry. Analysis based upon equations of respiratory gas exchange suggests that it is important to pay more attention to differences in the solubility of pollutants in different marine environments rather than assume some constant solubility. Solubility and diffusion coefficients of different pollutants in portions of the water/blood barrier also requires further investigation. The most readily available part of this pathway would appear to be the mucus covering the gill surface which increases in thickness under the action of many pollutants.

Allometric considerations suggest that smaller specimens of a given species would be more sensitive to pollutants than larger individuals.

Finally, it seems probable that pollutants in sublethal concentration would have more effect on the active oxygen consumption than the standard consumption. As a consequence the scope for aerobic activity of different fish species would be affected to a greater or lesser extent. Few studies have so far investigated the extent of pollutant action on this important aspect of a fish's life.

References

Baumgarten-Schumann D, Piper J (1968) Gas exchange in the gills of resting unanaesthetised dogfish *(Scyliorhinus stellaris)*. Respir Physiol 5: 317–325

Brown CE, Muir BS (1970) Analysis of ram ventilation of fish gills with application to skipjack tuna *(Katsuwonus pelamis)*. J Fish Res Board Can 27: 1637–1652

Brown VM (1976) Aspects of heavy metals toxicity in fresh waters. In: Andrew RW, Hodson PV and Konasewich DE (eds) Toxicity to biota of metal forms in natural water, chapter 3. Great Lakes International Joint Commission's Research Advisory Board. Windsor, Ontario, p 59–75

Daniel TL (1981) Fish mucus: in situ measurements of polymer drag reduction. Biol Bull Mar Biol Lab (Woods Hole) 160: 367–382

Dejours P (1973) Problems of control of breathing in fishes. In: Bolis L, Schmidt-Nielsen K and Maddrell SHP (eds) Comparative physiology. North-Holland, pp 117–133

Dejours P (1976) Water versus air as the respiratory media. In: Hughes GM (ed) Respiration of amphibious vertebrates. Academic, London, pp 1–15

Douglas EL, Friedel WA, Pickwell GV (1976) Fishes in oxygen-minimum zones: blood oxygenation characteristics. Science (Wash DC) 191: 957–959

Duthie GG, Hughes GM (1982) Some effects of gill damage on the swimming performance of rainbow trout *(Salmo gairdneri)*. J Physiol 327: 21P–22P

Fry Fej, Hart JS (1948). The relation of temperature to oxygen consumption in the goldfish. Biol Bull Mar Biol Lab (Woods Hole) 94: 66–77

Graham JB, Laurs RM (1982) Metabolic rate of the albacore tuna *Thunnus alalunga.* Mar Biol (NY) 72: 1–6

Hemmingsen AM (1960) Energy metabolism as related to body size and respiratory surfaces and its evolution. Reports of the Steno Mem. Hosp. and the Nordisk Insulinlab. IX, pp 1–110

Heusner AA (1983) Body size, energy metabolism and the lungs. J Appl Physiol 54: 867–873

Holeton GF (1980) Oxygen as an environmental factor of fishes. In: Ali MA (ed) Environmental physiology of fishes. Plenum, New York, pp 7–32

Holeton GF, Randall DJ (1967) The effect of hypoxia upon the partial pressure of gases in the blood and water afferent and efferent to the gills of rainbow trout. J Exp Biol 46: 317–327

Hoyt JW (1975) Hydrodynamic drag reduction due to fish slimes. In: Wu TY–T, Brokaw CJ and Brennen C (eds) Swimming and flying in nature, Vol 2. Plenum, London, pp 653–672

Hughes GM (1960) A comparative study of gill ventilation in marine teleosts. J Exp Biol 37: 28–45

Hughes GM (1961) How a fish extracts oxygen from water. New Sci 11: 346–348

Hughes GM (1963) Comparative physiology of vertebrate respiration, 2nd edn 1974. Heinemann London

Hughes GM (1964) Fish respiratory homeostasis. Symp Soc Exp Biol 18: 81–107

Hughes GM (1966a) The dimensions of fish gills in relation to their function. J Exp Biol 45: 177–195

Hughes GM (1966b) Evolution between air and water. In: de Reuck AVS and Porter R (eds) Ciba Foundation symposium on "Development of the Lung". Churchill, London, pp 64–80
Hughes GM (1967) Experiments on the respiration of the trigger fish *(Balistes capriscus)*. Experientia (Basel) 23: 1077
Hughes GM (1970) Morphological measurements on the gills of fishes in relation to their respiratory function. Folia Morphol (Prague) 18: 78–95
Hughes GM (1972) Morphometrics of fish gills. Respir Physiol 14: 1–25
Hughes GM (1973) Respiratory responses to hypoxia in fish. Am Zool 13: 475–489
Hughes GM (1976a) Polluted fish respiratory physiology. In: Lockwood APM (ed) Effects of pollutants on aquatic organisms. SEB seminar series 2. Cambridge University Press, pp 163–183
Hughes GM (1976b) On the respiration of *Latimeria chalumnae*. J Linn Soc (Lond) Zool 59: 195–208
Hughes GM (1977) Dimensions and the respiration of lower vertebrates. In: Pedley TJ (ed) Scale effects in animal locomotion. Academic, London, pp 57–81
Hughes GM (1979) Scanning electron microscopy of the respiratory surface of trout gills. J Zool (Lond) 188: 443–453
Hughes GM (1982) An introduction to the study of gills. In: Houlihan DF, Rankin JC and Shuttleworth TJ (eds) Gills. SEB seminar series 16. Cambridge University Press, pp 1–24
Hughes GM (1984a) Scaling of respiratory areas in relation to oxygen consumption of vertebrates. Experimentia (Basel) 40: 519–524
Hughes GM (1984b) General anatomy of the gills. In: Hoart WS and Randall DJ (eds) Fish physiologie, vol. XA. Academic, New York, pp 1-72
Hughes GM, Johnston IA (1978) Some responses of the electric ray *(Torpedo marmorata)* to low ambient oxygen tensions. J Exp Biol 73: 107–117
Hughes GM, Knights B (1968) The effect of loading the respiratory pumps on the oxygen consumption of *Callionymus lyra*. J Exp Biol 49: 603–615
Hughes GM, Perry SF (1976) Morphometric study of trout gills: a light microscopic method suitable for the evaluation of pollutant action. J Exp Biol 64: 447–460
Hughes GM, Shelton G (1962) Respiratory mechanisms and their nervous control in fish. In: Lowenstein O (ed) Advances in comparative physiology and biochemistry, vol I. Academic, London, pp 275–364
Hughes GM, Tort L (1983) Recovery of cardiac and ventilatory frequencies after zinc treatment. Abstract – 5th Conference of the European Society for comparative physiology and biochemistry, Taormina (Italy), pp 98–99
Hughes GM, Umezawa S-I (1968) On respiration in the dragonet *Callionymus lyra* (L.). J Exp Biol 49: 565–582
Hughes GM, Wright DE (1970) A comparative study of the ultrastructure of the water/blood pathway in the secondary lamellae of teleost and elasmobranch fishes – benthic forms. Z Zellforsch Mikrosk Anat 104: 478–493
Hughes GM, Kikuchi Y, Watari H (1982) A study of the deformability of red blood cells of a teleost fish, the yellowtail *(Seriola quinqueradiata)* and a comparison with human erythrocytes. J Exp Biol 96: 209–220
Hughes GM, Perry SF, Brown VM (1979) A morphometric study of effects of nickel, chromium and cadmium on the secondary lamellae of rainbow trout gills. Water Res 13: 665–679
Johansen K (1982) Respiratory gas exchange of vertebrate gills. In: Houlihan DF, Rankin JC and Shuttleworth TJ (eds) Gills. SEB seminar series 16. Cambridge University Press, pp 99–128
Krogh A (1919) The rate of diffusion of gases through animal tissue. J Physiol (Lond) 52: 391–408
Larsson A (1973) Clinic-chemical methods applied to fish blood with reference to effects of chlorinated hydrocarbons. In: Bolis L, Schmidt-Nielsen K ánd Maddrell SHP (eds) Comparative physiology. North Holland, pp 619–628
MacDonald AG (1972) The role of high hydrostatic pressure in the physiology of marine animals. Symp Soc Exp Biol 26: 209–232
Miller TG, Mackay WC (1982) Relationship of secreted mucus to copper and acid toxicity in rainbow trout. Bull Environ Contam Toxicol 28: 68–74

Pauly D (1981) The relationships between gill surface area and growth performance in fish: a generalisation of von Bertalanffy's theory of growth. Meeresforschung 28: 251–282

Rahn H (1966) Gas transport from the environment to the cell. In: de Reuck AVS and Porter R (eds) Ciba Foundation symposium on "Development of the Lung". Churchill, London, pp 3–23

Rahn and Howell BJ (1976) Bimodal gas exchange. In: Hughes GM (ed) Respiration of amphibious vertebrates. Academic, London, pp 271–285

Randall DJ (1970) Gas exchange in fish. In: Hoar WS and Randall DJ (eds) Fish physiology, vol 4. Academic, New York, pp 252–292

Rayner MD, Keenan MJ (1967) Role of red and white muscles in the swimming of the skipjack tuna. Nature (Lond) 214: 302–303

Roberts JL (1975) Active branchial and ram gill ventilation in fishes. Biol Bull Mar Biol Lab (Woods Hole) 148: 85–105

Roberts JL, Graham JB (1979) Effect of swimming speed on the excess temperatures and activities of heart and red and white muscles in the mackerel *Scomber japonicus.* Fish Bull SAfr 76: 861–867

Saunders RL (1962) The irrigation of the gills in fishes. II. Efficiency of oxygen uptake in relation to respiratory flow, activity, and concentrations of oxygen and carbon dioxide. Can J Zool 40: 817–862

Schmidt-Nielsen K (1970) Energy metabolism, body size and problems of scaling. Federation Proceedings 29: 1524–1531

Shelton G (1970) The regulation of breathing. In: Hoar WS and Randall DJ (eds) Fish physiology, vol 4. Academic, New York, pp 293–459

Singh BN (1976) Balance between aquatic and aerial respiration. In: Hughes GM (ed) Respiration of amphibious vertebrates. Academic, London, pp 125–164

Skidmore JF, Tovell PWA (1972) Toxic effects of zinc sulphate on the gills of rainbow trout. Water Res 6: 217–230

Stevens ED (1972) Some aspects of gas exchange in tuna. J Exp Biol 56: 809–823

Stevens ED, Randall DJ (1967) Changes of gas concentrations in blood and water during moderate swimming activity in rainbow trout. J Exp Biol 46: 329–337

Winberg GG (1956) (1) Rate of metabolism and food requirements of fishes. Fish Res Board Can Translation series No 194

Transfer Changes in Fish Gills During Stress

A.-L. UNGELL, A. KIESSLING, and S. NILSSON[1]

1 Introduction

Marine fish live in a complex, although for most species rather stable environment. However, changes in pH, salinity, and oxygen availability are of concern for the migratory and estuarine fish and pollutants may affect marine fish populations exposed to industrial waste water. In these situations the fish may be exposed to various types of "stress" (stress stimuli) such as hypoxia, physical damage, changes in temperature and salinity or, in some cases, pollutants (see Fig. 1). Such stimuli produce physiological changes in the animal (stress response), including an increased sympathico-adrenal activity which leads to increased plasma levels of catecholamines and also release of corticosteroids (Fig. 1). The stress response induced by short-term exposure to stress includes the various types of defence reactions (cf. Cannon 1929), which allow the animal maintain homeostasis in the new environment, or to move away from the stressful stimulus.

In this paper, we will discuss very briefly the role of the adrenergic system in the stress response, with particular emphasis on the control of gill functions.

1.1 Stress

The term stress may be used for an environmental stimulus (stressor) which triggers a physiological response (stress response) in the organism (cf. Pickering 1981). This overall response can be divided into primary effects (central or local), and secondary effects resulting from the central primary effects (Fig. 1).

Pollutants form a special group of stress stimuli, sometimes with very severe effects on the organism. Such substances will directly affect the gills, which expose a large surface area to the water, and damage of the gills may occur, e.g., due to precipitation of the pollutant on the gill surfaces (Herbert et al. 1961; Hughes 1981). Furthermore, pollutants entering the blood through the gill membranes can affect all parts of the fish (Förlin et al. 1983).

1 Department of Zoophysiology, University of Göteborg, S–400 31 Göteborg, Sweden

Toxins, Drugs, and Pollutants in Marine Animals
(ed. by Bolis et al.)

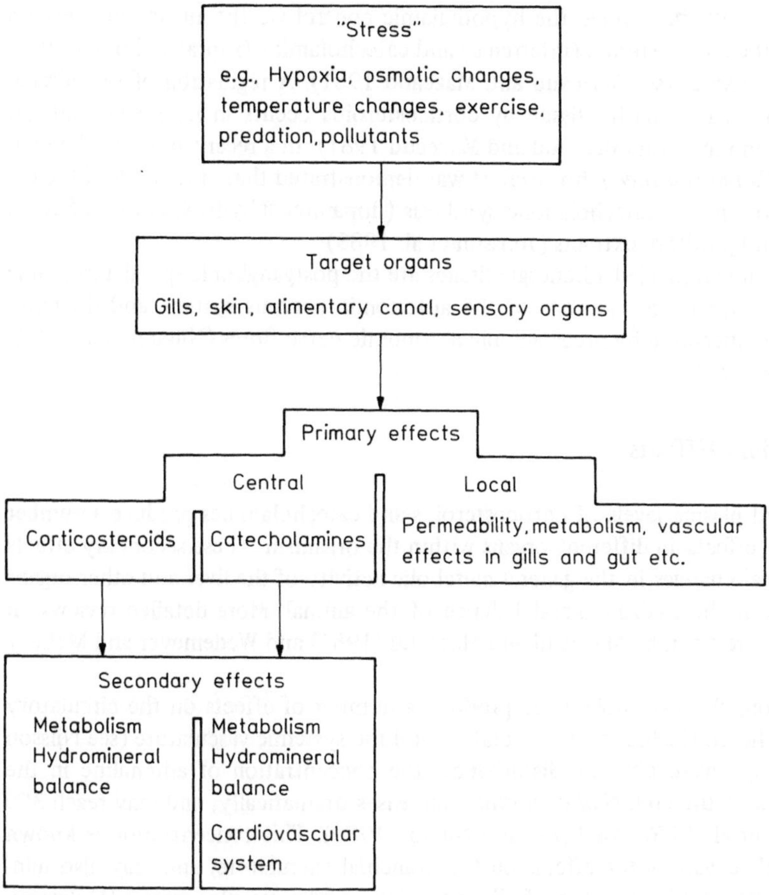

Fig. 1. A summary of the effects of stress in fish. (For further details and references, see Mazeaud and Mazeaud 1981)

1.2 Target Organs

The targets of the stress stimuli are the tissues which are directly exposed to the environmental changes: the gills, skin, alimentary canal and various sensory organs. Both the gills and the gut have large surface areas and a high permeability to different substances, and are therefore important sites of entry for, e.g., pollutants (Haywood et al. 1977; Isaia et al. 1978; Girard and Payan 1980).

1.3 Primary Effects

The primary effects induced by stress can be either local or central. One example of a local effect is the general increase in branchial vascular resistance induced by hypoxia, a phenomenon which will increase the efficiency for the oxygen uptake in the gills in a situation with a low oxygen tension in the water (for references see Petterson 1983).

The central effects involve the hypothalamic control via the autonomic nervous system and the corticosteroid (interrenal) and catecholamine (suprarenal or chromaffin) secreting tissues (see Mazeaud and Mazeaud 1981). A regulation of the adrenaline synthesis in chromaffin tissue by corticosteroids occurs in mammals, but has been denied in most fish (Mazeaud and Mazeaud 1981). In a recent study of the rainbow trout *(Salmo gairdneri)*, however, it was demonstrated that at least on of the enzymes involved in the catecholamine synthesis (dopamine-β-hydroxylase) is affected by cortisol and pituitary extracts (Jönsson et al. 1983).

The centrally controlled adrenergic tissues are the postganglionic spinal autonomic (sympathetic) adrenergic neurons of the autonomic nervous system, and the chromaffin tissue innervated by preganglionic autonomic nerve fibres (Nilsson et al. 1975, 1976; Nilsson 1983).

1.4 Secondary Effects

The increased plasma levels of corticosteroids and catecholamines produce a number of secondary effects in different organs within the organism. These secondary effects include radical changes in the general metabolic activity of the liver and other organs and changes in the hydro-mineral balance of the animal. More detailed reviews on these effects are given by Mazeaud and Mazeaud (1981) and Wedemeyer and McLeay (1981).

In addition, the catecholamines produce a number of effects on the circulatory system: the heart, the branchial vasculature and the systemic vasculature (see Nilsson 1983). During severe physical disturbance, the concentration of adrenaline in the blood plasma of the cod, *Gadus morhua*, increases dramatically, and may reach 300 nM (Nilsson et al. 1976; Wahlqvist and Nilsson 1980). This concentration is known to have marked vasomotor effects on the branchial vasculature, and may also reinforce the adrenergic innervation of all parts of the cardio-vascular system (Holmgren 1977; Wahlqvist 1980, 1981; Nilsson 1983).

In the gills, the vasculature may thus, during stress, be affected by

a) a direct (primary) effect of stress, such as the local vasoconstriction induced by hypoxia (see Sec. 1.3)
b) increased levels of circulating catecholamines which will induce an overall decrease in the branchial vascular resistance and increase the efficiency of oxygen uptake (Wahlqvist 1980; Pettersson 1983)
c) increased adrenergic nerve activity, which will induce an overall increase in the branchial vascular resistance and a redistribution of the branchial blood flow (Nilsson and Pettersson 1981)

In addition, direct effects of catecholamines on the gill membranes affect ion transfer and possibly also the permeability of other substances (see Sect. 2 below).

2 Gill Transfer During Stress

Increased plasma levels of catecholamines or increased adrenergic nerve activity will produce a redistribution of the blood flow within the gills, and this phenomenon will increase the functional surface area of the gills and thus the transfer of various substances (oxygen, water, urea etc) (Fig. 2) (Bergman et al. 1974; Pettersson 1983). The vascular adjustments associated with catecholamines are thought to be either an increase in the number of secondary lamellae perfused at any given moment (lamellar recruitment model; Hughes 1972; Booth 1978, 1979), or a redistribution of blood from the nonrespiratory basal channels of the secondary lamellae to the central parts of the lamellae which are in more intimate contact with the surrounding water (Farrell et al. 1979, 1980).

An increase in the specific transfer across the gills of a certain substance could also be induced by a modification of the gill membranes resulting in an increased permeability to that substance (Fig. 2) (cf. Haywood et al. 1977; Isaia et al. 1978).

In the very preliminary experiments described below, an attempt has been made to separate the vascular effects of adrenaline from its possible direct effects on the permeability of the gill membranes.

2.1 Effects of Adrenaline on the Transfer of Oxygen and ^{14}C-Urea in isolated Gills

In these experiments isolated gill arches from the Atlantic cod, *Gadus morhua*, were perfused with a Ringer's solution (pO_2 = 35 ± 10 mmHg) from a pulsatile pump with independently adjustable stroke volume and frequency. The gill arch was submerged in recirculating sea water at 10°C (pO_2 = 150 ± 5 mmHg) rapidly flowing past the gill arch. The transfer of oxygen and ^{14}C-urea from the sea water to the perfusion fluid was monitored, together with the inflow (afferent) pressure (P_v) and the efferent arterial outflow ($\dot{Q}$). The efferent arterial back pressure (P_d) was kept at 2.0 kPa, and the branchial vascular resistance estimated as $(P_v - P_d)/Q$.

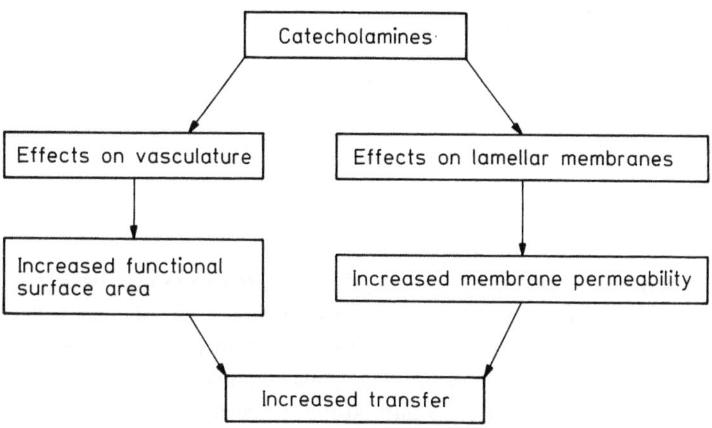

Fig. 2. A summary of two possible pathways for the catecholamine-induced increase in gill transfer

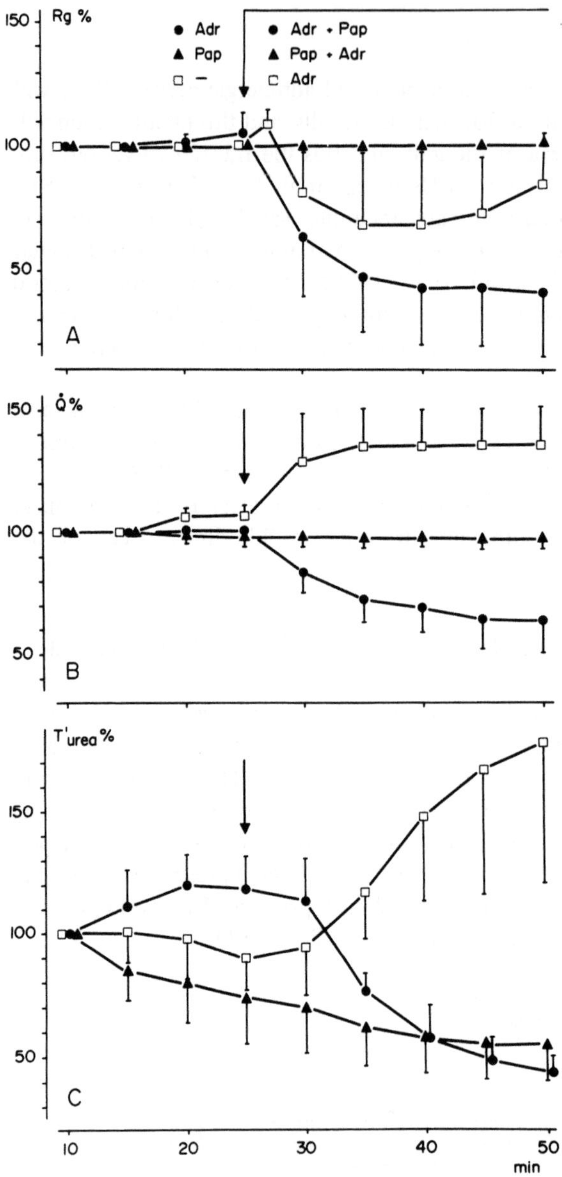

Fig. 3 A–C. Percentual changes in **A** the branchial vascular resistance (R_g); **B** the efferent arterial flow ($\dot{Q}$) and **C** the transfer factor for ^{14}C-urea (T'_{urea}) during perfusion of isolated gill arches from the cod, *Gadus morhua*. Additions of drugs were made as indicated at the top of the figure: *Adr* adrenaline 10^{-6} M; *Pap* papaverin 10^{-4} M. Note that the three different curves of each diagram have been normalized to a common starting point (= 100%), and that the absolute changes therefore cannot be compared. Each *point* shows the mean value of 4–8 experiments and *vertical bars* show the standard error of the mean (S.E.M.)

Three types of experiment were performed. In the first group the gills were perfused with pure Ringers's solution, and after 25 min adrenaline (10^{-6}M) was added (Fig. 3). In the second group adrenaline (10^{-6}M) was present in the perfusion fluid from the start of the experiment, and after 25 min papaverin (10^{-4}M) was added to paralyze the vascular smooth muscle. In the third group papaverin (10^{-4}M) was present from the start, and adrenaline (10^{-6}M) added after 25 min (Fig. 3).

The percentual effects of the various drug combinations on the branchial vascular resistance (R_g), efferent arterial flow ($\dot{Q}$) and transfer factor for urea (T'_{urea}) are summarized in Fig. 3. Adrenaline produced a biphasic change in the branchial vascular resistance (Fig. 3A) due to the α-adrenoceptor mediated vasconstriction followed by the β-adrenoceptor mediated vasodilatation demonstrated in numerous teleost species (for references see, e.g., Nilsson 1984). An increase in the efferent arterial flow was also evident after adrenaline (Fig. 3B), and this effect is probably due to a reduction in the branchial venous and/or efferent arterial leakage from the preparation (cf. Nilsson and Pettersson 1981). Addition of papaverin (10^{-4}M) to the adrenaline-containing perfusion fluid produced paralysis of the smooth muscle of the branchial vasculature, manifested as a decrease in both the branchial vascular resistance and the efferent arterial flow (Fig. 3 AB). If papaverin is present in the perfusion fluid from the start of the experiment, no vasomotor effects of adrenaline can be recorded (Fig. 3 AB).

The transfer factors for urea (Fig. 3C) and oxygen (not shown) were maintained at a steady level, or increase slightly if adrenaline are present in the perfusion fluid from the start of the perfusion. In the absence of adrenaline there was a slow but steady decline in the T'_{urea}, which, at least during the early stages of a perfusion, could be reversed by adrenaline. The decline in T'_{urea} was similar or slightly more pronounced in the papaverin-treated preparations, and in this case no effect of adrenaline could be demonstrated (Fig. 3C). The changes in oxygen transfer followed a similar pattern but the results are less clear-cut, probably due to the oxygen consumption of the preparation.

The results show that recordable changes in ^{14}C-urea transfer caused by adrenaline is solely due to the papaverin-sensitive, probably vasomotor, effects of adrenaline. The study also suggests that papaverin, or functionally related drugs, may prove important chemical tools which can be utilized to differentiate between the vasomotor component and the possible direct membrane permeability component controlling the transfer of compounds such as gases, urea, water etc.

2.2 Epithelial Detachment as a Limiting Factor for Gill Transfer

The decline in T'_{urea} observed in preparations perfused in the absence of adrenaline was investigated further in the light of the findings that isolated gill preparations from freshwater fish perfused with Ringer's solutions showed a detachment of the branchial epithelium (branchial edema) which severely impaired the transfer of oxygen and other compounds across the gills of rainbow trout *(Salmo gairdneri*; Pärt et al. 1982) and eel (*Anguilla australis*; Ellis and Smith 1983). The structural deterioration of the gills could be prevented by the presence of adrenaline in the perfusion fluid, or by perfusion with whole blood or blood plasma. In such experiments a high and steady oxygen transfer is maintained during the experiment (Pärt et al. 1982; Ellis and Smith 1983).

From our own histological studies of cod gills perfused with and without adrenaline, it is evident that a profound epithelial detachment takes place in gills perfused in the absence of adrenaline (Fig. 4 AB). Our results thus corroborate those of Pärt et

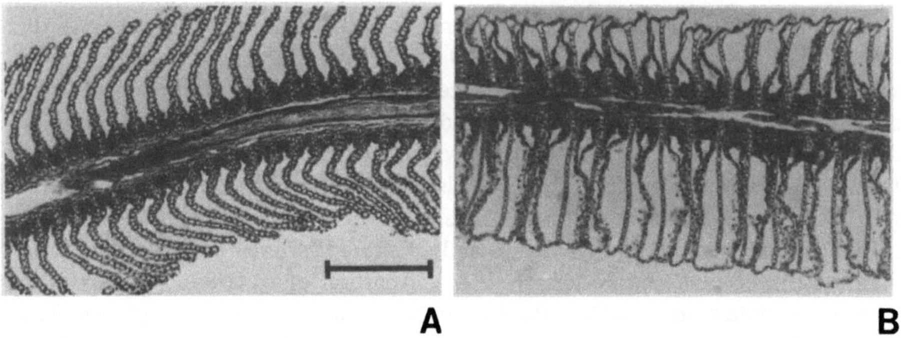

Fig. 4 A, B. Transverse sections through the secondary lamellae of cod gills. **A** a gill perfused with adrenaline (10^{-6} M) present in the perfusion fluid. The pO_2 of the efferent arterial outflow from the gill arch was approximately 85% of that in the external sea water. **B** a gill perfused without adrenaline in the perfusion fluid. In this case the pO_2 in the efferent arterial outflow was approximately 30% of that in the external sea water. Note the very marked epithelial detachment. Calibration bar = 200 μm

al. (1982) and Ellis and Smith (1983) and demonstrate the importance of carefully considering the possible structural changes in the gills in all studies dealing with measurements of gill transfer both in fresh water and sea water teleosts. The demonstration of the development of a diffusion barrier in the gills during experimental conditions may necessitate re-interpretation of several older studies of transfer in fish gills.

References

Bergman HL, Olson KR, Fromm PO (1974) The effects of vasoactive agents on the functional surface area of isolated perfused gills of rainbow trout. J Comp Physiol 94: 267–286

Booth JH (1978) The distribution of blood in the gills of fish: application of a new technique to rainbow trout *(Salmo gairdneri)*. J Exp Biol 73: 119–129

Booth JH (1979) The effects of oxygen supply, epinephrine, and acetylcholine on the distribution of blood flow in trout gills. J Exp Biol 83: 31–39

Cannon WB (1929) Organization for physiological homeostatis. Physiol Rev 9: 399–431

Ellis AG, Smith DG (1983) Oedema formation and impaired O_2 transfer in Ringer perfused gills of the eel *Anguilla australis*. J Exp Zool 227: 371–380

Farrell AP, Daxboeck D, Randall DJ (1979) The effect of input pressure and flow on the pattern and resistance to flow in the isolated perfused gill of a teleost fish. J Comp Physiol 133: 233–240

Farrell AP, Sobin SS, Randall DJ, Crosby S (1980)I Intralamellar blood flow patterns in fish gills. Am J Physiol 239: R428–R436

Förlin L, Andersson T, Koivusaari U, Hansson T (tobe published 1983) Influence of biological and environmental factors on hepatic steroid and xenobiotic metabolism in fish: interaction with PCB and betanaphtoflavone. Mar Environ Res

Girard JP, Payan P (1980) Ion exchanges through respiratory and chloride cells in feswater- and seawater adapted teleosteans. Am J Physiol 238: R260–R268

Haywood GP, Isaia J, Maetz J (1977) Epinephrine effects on branchial water and urea flux in rainbow trout. Am J Physiol 232: R110–R115

Herbert DWM, Alabaster JS, Dart MC, Lloyd R (1961) The effect of china-clay wastes on trout streams. Int J Air Wat Pollut 5: 56–74

Holmgren S (1977) Regulation of the heart of a teleost *Gadus morhua* by autonomic nerves and circulating catecholamines. Acta Physiol Scand 99: 62–74

Hughes GM (1972) Morphometrics of fish gills. Respir Physiol 14: 1–25

Hughes GM (1981) Effects of low oxygen and pollution on the respiratory system of fish. In: Pickering AD (ed) Stress and Fish. Academic, London, pp 121–146

Isaia J, Maetz J, Haywood GP (1978) Effects of epinephrine on branchial non-electrolyte permeability in rainbow trout. J Exp Biol 74: 227–237

Jönsson A-C, Wahlqvist I, Hansson T (1983) Effects of hypophysectomy and cortisol on the catecholamine biosynthesis and catecholamine content in chromaffin tissue from rainbow trout *Salmo gairdneri.* Gen Comp Endocrinol 51: 278–285

Mazeaud MM, Mazeaud F (1981) Adrenergic responses to stress in fish. In: Pickering AD (ed) Stress and fish. Academic, New York, pp 49–75

Nilsson S (1983) Autonomic nerve function in the vertebrates. Springer, Berlin Heidelberg New York

Nilsson S (to be published 1984) Innervation and pharmacology of the gills. In: Hoar WS and Randall DJ (ed) Fish physiology, vol X. Academic New York

Nilsson S, Pettersson K (1981) Sympathetic nervous control of blood flow in the gills of the Atlantic cod *Gadus morhua.* J Comp Physiol 144: 157–163

Nilsson S, Holmgren S, Grove DJ (1975) Effects of drugs and nerve stimulation on the spleen and arteries of two species of dogfish *Scyliorhinus canicula* and *Squalus acanthias.* Acta Physiol Scand 95: 219–230

Nilsson S, Abrahamsson T, Grove DJ (1976) Sympathetic nervous control of adrenaline release from the head kidney of the cod *Gadus morhua.* Comp Biochem Physiol 55C: 123–127

Pärt P, Tuurala H, Soivio A (1982) Oxygen transfer, gill resistance and structural changes in rainbow trout *Salmo gairdneri* (Richardson) gills perfused with vasoactive agents. Comp Biochem Physiol 71C; 7–13

Pettersson K (1983) Adrenergic control of oxygen transfer in perfused gills of the cod *Gadus morhua.* J Exp Biol 102: 327–335

Pickering AD (1981) Introduction: The concept of biological stress. In: Pickering AD (ed) Stress and fish. Academic, London, pp 1–9

Wahlqvist I (1980) Effects of catecholamines on isolated systemic and branchial vascular beds of the cod *Gadus morhua.* J Comp Physiol 137: 139–143

Wahlqvist I (1981) Branchial vascular effects of catecholamines released from the head kidney of the Atlantic cod *Gadus morhua.* Mol Physiol 1: 235–241

Wahlqvist I, Nilsson S (1980) Adrenergic control ot the cardio-vascular system of the Atlantic cod *Gadus morhua* during 'stress'. J Comp Physiol 137: 145–150

Wedemeyer GA, McLeay (1981) Methods for determining the tolerance of fishes to environmental stressors. In: Pickering AD (ed) Stress and fish. Academic, London, pp 247–275

Effects of Acid Stress on Fish Gills

C.L. BOLIS[1], A. CAMBRIA[2], and M. FAMA[2]

1 Introduction

It is well known that the progressive acidification of the lakes in Canada as well as United States and Scandinavia, has reduced drammatically the fish production in these natural areas. This phenomenon, due to acid rain, could lead to very low pH values ranging between 4.5 and 4.0. Similar decrease in pH (4.7) has been found in several rivers of Nova Scotia, once teeming with salmon, and in other remote areas of Muskoka (Ontario). Moreover, water acidification may occur even in spring, when a large amount of acid water reaches the lakes after the melting of the snow. The pH then drops to 5.5, which is particularly harmful to small fish.

In this study we present some biochemical evidence by which the environmental acidification may determine alterations in aquatic organism. Since the gill membranes of fish are directly affected by external pH variations, it seems likely that alterations in their molecular structure may lead to a loss of function by their respiratory exchange mechanisms and osmotic regulation.

We undertook to research the relationship between external pH changes and phospholipid composition, as well as fatty acid distribution in phospholipid of gill membranes of *Salmo trutta.* For this purpose we reproduced the environmental conditions of low pH. In our previous experiments we found an high amount of plasmalogens in gill membranes confirming the observation of Dumont (1958) in *Eriocheir sinensis.* The plasmalogens are alkenil-ether phospholipids, similar to ester phospholipids, extremely sensitive to acid pH. Acid hydrolysis of such compounds yields/ both the monoacyl derivative and long-chain aliphatic aldehyde. These latter molecules show a detergent-like effect on lipid bilayer structure of membrane (Schulman 1967), pers. commun.).

2 Materials and Methods

Animals. Trout obtained from stream were maintained in a vat with fresh water continously filtered. Temperature, oxygen concentration, and pH were kept constant. The

1 Institute of General Physiology, University of Messina, Italy, Via dei Verdi, 85 – 98100 Messina, Italy
2 Institute of Biological Chemistry, University of Catania, V. le A. Doria, 6 – 95100 Catania, Italy

Toxins, Drugs, and Pollutants in Marine Animals
(ed. by Bolis et al.)

volume of the water was 40 l kg^{-1} of trout body weight. The pH was changed by slowly adding H_2SO_4 10 N from a buret. The pH value were constantly recorded an the lowest value (4.5 – 4.0) was reached after 36 h. These latter values were kept constant by adding small aliquots of H_2SO_4 10 N for 100–120 h.

Isolation and Purification of Lipids. At the end of this operation the trouts were killed by bleeding and the gills/were quickly removed and immediately frozen. The mucosa was repeatedly thawed and frozen to eliminate red cells and then homogenized with chloroform-metanol 2:1 (v/v). The lipids were extracted according to Folch et al. (1957). To prevent autoxidation in preparative work, 2,6-di-ter-butyl-p-cresol (50 mg/100 ml) was added to the solvent. The extracted lipids were separated into free fatty acids, neutral lipids, and phospholipids by silicic acid column chromatography (Dittmer and Wells 1969). The neutral lipids were eluted with chloroform and phospholipids with chloroform-metanol 1:4 (v/v).

Analysis. The neutral lipids were metanolized (0.5 M anhydrous methanolic HCl, 24 h, 70°C) and the resulting fatty acids methyl esters were extracted with hexane. The analysis was performed with a Fractovap Model 4200 gas chromatograph (C. Erba, Milan. Italy) equipped with dual flame ionization detectors and with electronic programmer. The columns were 3% SE 30 on Chromosorb WHP 80–100 mesh. The pressure of carrier gas (N_2) was 1.2 kg cm^{-2}; the temperature 220°C. The percentage of various fatty acids were determined by calculating the peak areas by triangulation. The fatty acids methyl esters were identified by comparing the relative retention times/with those of known standards.

The phospholipid extract was taken up with chloroform and the phosphate content was assayed according to Bartlett's method (1959). The phospholipids were separated from each other by two-dimensional thin layer chromatography (TLC) on silica gel F_{254} plate (Merk, Darmstadt), by using as developing solvent: first direction, chloroform/methanol/NH_4OH 7 N (60:30:5); second direction, chloroform/ methanol/glacial acetic acid/water (65:43:1:3). After development, the plates were sprayed with 0.2% 2,7-dichlorofluorescein in ethanol (w/v) and spots were outlined under UV light and scraped off. The phospholipids were eluted from the gel, using chloroform:methanol:water (65:25:4). Each fraction was analyzed for inorganic phosphate according to micromethod of Hess and Derr (1975). Plasmalogens content was instead determined according to Carey's method (1982) based on Hg binding with the alkenyl group. The chromatographic techniques of Owens (1966) were used to identify the lysoderivaties of plasmalogens, choline and ethanolamine. The relative spots were eluted and inorganic phosphate was determined by the micromethod already described.

3 Results

The percentual distribution of different phospholipid classes in the gill membranes of *Salmo trutta,* is shown in Table 1.

For the ester phospholipids the major fractions are represented, in order to decrease concentration, from phosphatidylcholine (PC) phosphatidilethanolamine (PE), sphyngomielin (SP), phosphatidylserine (PS) and phosphatidylinositol (PI).

Table 1. Phospholipid distribution in gill membranes of *Salmo trutta* exposed to low pH ranges

	Phospholipid					Plasmalogens		Lisoplasmalogens	
Control	PE	PC	SP	PI	PS	PE	PC	LPE	LPC
	13.6 ± 0.5	23.9 ± 1.2	12.3 ± 0.8	3.8 ± 0.2	8.5 ± 0.6	7.6 ± 0.8	20.5 ± 1.4	0.8 ± 0.07	1.5 ± 0.2
Acid water	12.5 ± 0.7	24.4 ± 0.9	11.8 ± 0.5	2.7 ± 0.4	7.6 ± 0.4	5.2 ± 0.4	26.0 ± 0.8	1.7 ± 0.2	4.5 ± 0.3

The fish were maintained in aquarium at pH between 4.5 and 4.0 for 100 – 120 h. Values are expressed in percentage (%) total lipid phosphorus. Means ± S.D. for three separate experiments.

Table 2. *Salmo trutta* gills. Percentage fatty acid composition in total phospholipids[a]

Fatty acid	Normal	% Unsaturate	Acid water	% Unsaturate
14:0	5.87	42.73	1.47	70.98
16:0	26.75		7.67	
16:1	10.89		1.38	
18:0	15.79		15.27	
18:1	28.32		12.03	
20:0	8.83		2.08	
20 unsat.	trace		25.72	
22:0	trace		0.43	
22 unsat.	3.51		27.01	

[a] Values are expressed as weight percentage of fatty acid methyl esters. Data represent means of 3 to 5 determinations that were run in duplicate

There are also present two classes of alkenyl-ether phospholipids (plasmalogen) of both choline and ethanolamine, respectively 46.1% and 35.8% with respect to the ester phospholipids. Morever it can be observed a small percentage of plasmalogen lysoderivatives.

The exposition of trouts to acid environment (pH = 4.5 – 4.0) during the experiment (100–120 h) does not determine detectable modification in the distribution of different phospholipid species of the gills, unless for the lysoderivatives of plasmalogens (2-acyl lysophosphoderivatives) whose percent concentration increases remarkably.

The percent composition of fatty acids in gill total phospholipids is shown in Table 2. As can be seen, in the phospholipids of normal gills there is a net predominance of C_{16} and C_{18} fatty acids, whereas those with higher carbon-chain are present either in small quantities or in traces.

The stay in acid water determines a decrease of the above mentioned fatty acids, especially those with a C_{16} chain, and a pronounced increase of C_{20} and C_{22} unsatured fatty acids (+ 66%). These results are clearly demonstrated in the chromatograms of Figs. 1 and 2. In Table 3 it is reported the fatty acid composition of gill neutral lipids. It is possible to observe, rather than in the phospholipids, a predominance of long-chain fatty acids, C_{20} and C_{22} both saturated and unsaturated, which represent the 64% of the total. The acidification of the water does not noticeably modify the kind of distribution in the different molecular species, but it increases the percent of insaturation (+ 28%), due to C_{20} and C_{22} not saturated.

Figs. 3 and 4 give typical chromatograms, clearly showing the results previously mentioned.

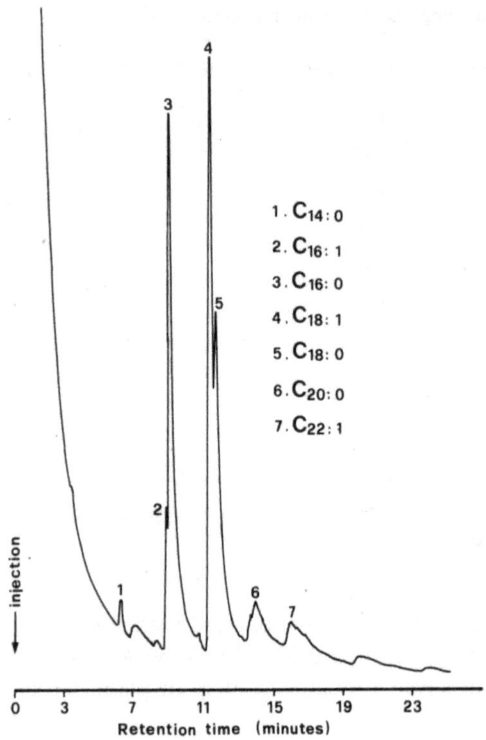

Fig. 1. Chromatogram of fatty acid methyl esters obtained from total phospholipids of gill membranes of normal *Salmo trutta*. Experimental conditions are reported in Material and Methods

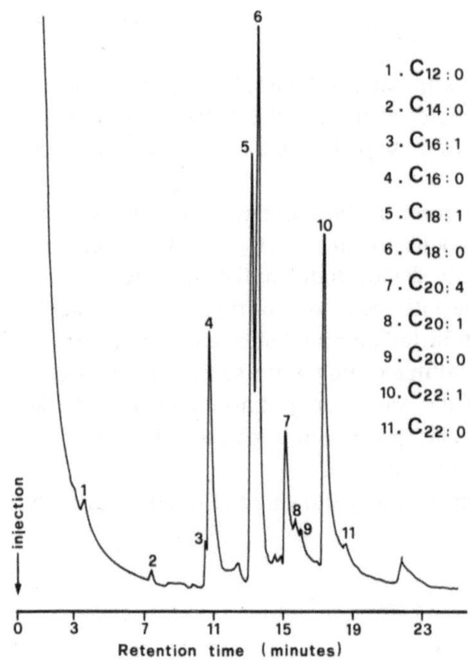

Fig. 2. Chromatogram of fatty acid methyl esters obtained from total phospholipids of gill membranes of **Salmo trutta** maintained in acid water. Experimental conditions are reported in Materials and Methods

Table 3. *Salmo trutta* gills. Percentage fatty acid composition in neutral lipids[a]

Fatty acid	Normal	% Unsaturate	Acid water	% Unsaturate
12:0	Trace	39.25	Trace	50.37
14:0	9.62		15.35	
15:0	3.61		0.90	
16:0	7.68		3.72	
16:1	2.08		1.28	
17:0	Trace		Trace	
17:1	1.04		0.65	
18:0	5.03		1.97	
18 unsat.	6.39		2.32	
20:0	34.75		25.59	
20 unsat.	11.99		15.51	
22 unsat.	16.64		28.48	

[a] Values are expressed as weight percentage of fatty acid methyl esters. Data represent means of 3 to 5 determinations that were run in duplicate

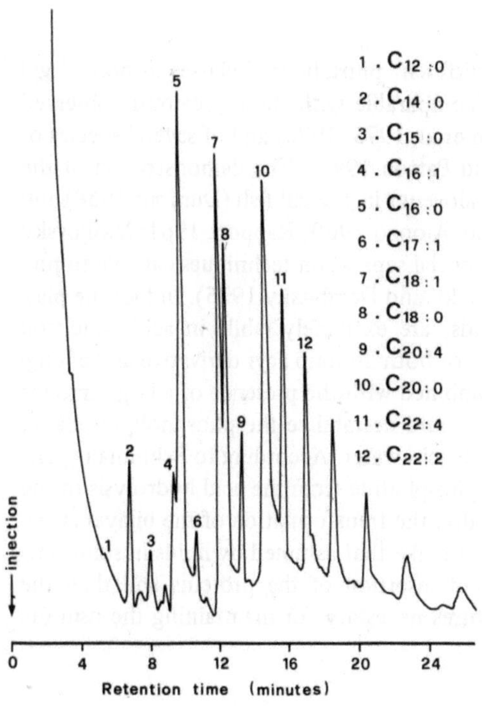

Fig. 3. Chromatogram of fatty acid methyl esters obtained from neutral lipids of gill membranes of normal *Salmo trutta*. Experimental conditions are reported in Materials and Methods

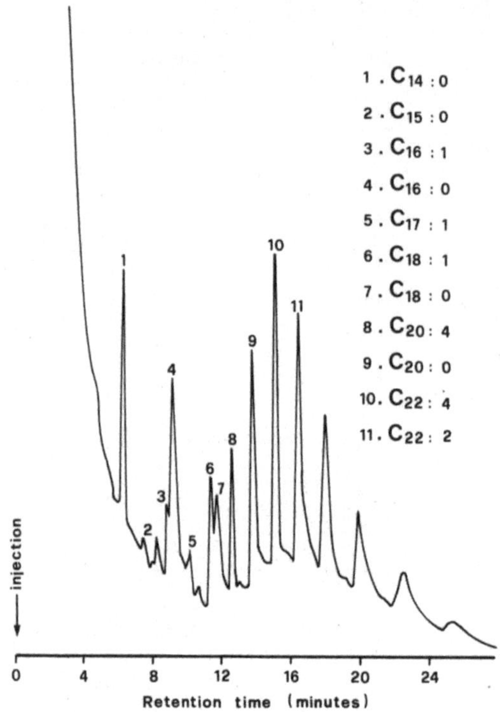

Fig. 4. Chromatogram of fatty acid methyl esters obtained from neutral lipids of gill membranes of *Salmo trutta* maintained in acid water. Experimental conditions are reported in Materials and Methods

4 Discussion

As far as they concern the percent of different phospholipid classes in normal gill of *Salmo trutta,* the values obtained are comparable with those previously observed in the gill of fresh water eel (Zwingelstein et al. 1973, 1975) and of several species of marine fish (Anderson 1970; Thomas and Patton 1972). The demonstration of the presence of alkenyl-phospholipids (plasmalogens) in the gill fish (Dumont 1958) and in the marine invertebrates (Rapport and Alonzo 1960; Rapport 1961; Vaskovsky 1969) was possible because of the use of special separation techniques capable to prevent the hydrolysis (Owens 1966; Vaskovsky and Dembitsky 1975). In fact the plasmalogens, rather than ester phospholipids, are extremely labile in acid condition where they hydrolyze with the formation of both 2-mono acyl derivative and a long-chain aldehyde. This streaking feature, combined with the presence of a large amount of such molecules in the gill membranes, could destabilize the phospholipid bilayer structure when this one is in contact with a acid water. According to Schulman (pers. commun. 1967), the formation of 2-acyl-phosphatide from the acid hydrolysis of the plasmalogen, induces, completely or partially, the transformation of the bilayer structure of the membrane, to a micellar structure like that assumed by lipids in solution.

This structure does not allow a correct insertion of the proteins and then the membrane would lose its functional features necessary for maintaining the osmotic equilibrium.

Moreover, during the acid stress, the increase of the unsaturated fatty acids content in the phospholipids, because of the presence of double bonds in cis-configuration, may influence the hydrophobic interactions and then the fluidity of the membrane. Such modifications have been observed by Chapelle and Pequeux (1983) in relation to different salt concentration of the environment.

Our results suggest that the alterations in both the phospholipid and fatty acid composition in *Salmo trutta* gill membranes during acid stress, may determine the damage of gill membrane structure, such to involve the loss of osmotic equilibrium regulation and of the gas exchange, with consequent death of the organisms.

References

Anderson TR (1970) Temperature adaptation and the phospholipids of membranes in gold fish (Carassius auratus). Comp Biochem Physiol 33: 663–687

Barlett GR (1959) Phosphorus assay in column chromatography. J Biol Chem 234: 466–468

Carey EM (1982) The quantitative determination of plasmalogens by its reaction with mercuric chloride. Lipids 17: 656–661

Chapelle S, Pequeux A (1983) Phospholipids and fluidity of anterior and posterior gills from chinese crab *Eriocheir sinensis* as related to the salinity of the environment. Symposium of 5th Conference of the European Society for Comparative Physiology and Biochemistry, Taormina, 5–8 September, pp. 130–131

Dittmer TC, Wells MA (1969) Quantitative and qualitative analysis of lipids and lipid components. In: Lowenstein JM (ed) Methods in enzymology, vol 14. Academic, New York, P 482

Dumont P (1958) Presence d'un acetalphosphatide dans les branchies du crabe *Eriocheir sinensis* en relation avec le mecanisme de transport actif des ions mineraux. Arch Int Physiol Biochim 66: 373–385

Folch J, Lees M, Sloane-Stanley GH (1957) A simple method for the isolation and purification of total lipids from animal tissues. J Biol Chem 226: 497–509

Hess HH, Derr TE (1975) Assay of inorganic and organic phosphorus in the 0.1–5 nanomole range. Anal Biochem 63: 607–613

Owens K (1966) A two dimensional thin-layer chromatographic procedure for the extimation of plasmalogens. J Biol Chem 100: 354–361

Rapport MM, Alonzo NF (1960) The structure of plasmalogens. V. Lipids of marine invertebrates. J Biol Chem 235: 1953–1956

Rapport MM (1961) The α, β-unsaturated ether (plasmalogen) content of the tissues of several mollusks. Biol Bull (Woods Hole) 121: 376–377

Thomas A, Patton S (1972) Phospholipids of fish gills. Lipids 7: 76–78

Vaskovsky VE, Kostetsky EY (1969) Phospholipids of marine invertebrates. Chem Phys Lipids 3: 102–105

Vaskovsky VE, Dembitsky VM (1975) Determination of plasmalogens contents in phospholipid classes by reaction micro-thin-layer chromatography. J Chromatogr 115: 645–649

Zwingelstein G, Meister R, Jouanneteau J (1973) Metabolisme des phospholipids de la branchie et du foie de l'anguille. Biochim (Paris) 55: 1495–1497

Zwingelstein G, Meister R, Brichon G (1975) Metabolisme comparé des phospholipides des organes effecteurs de l'osmoregulation chez l'anguille européenne (Anguilla anguilla). Biochimie (Paris) 57: 609–622

Superoxide Dismutase and Reduced Glutathione: Possible Defenses Operating in Hyperoxic Swimbladder of Fish

V. CALABRESE[1], F. GUERRERA[2], M. AVITABILE[1], M. FAMA[1], and V. RIZZA[1]

1 Introduction

Swimbladders of fish represent an example of gaseous environment in which the oxygen pressure significantly exceeds 1 bar (Haldane and Priestly 1935). While exposure of whole fish to oxygen partial pressures as great as their swimbladders results in oxygen poisoning, swimbladders appear to be insensitive to oxygen toxicity.

According to the "free radical theory" of oxygen toxicity (Gerschman et al. 1954), exposure to hyperoxia leads to elevated intracellular concentrations of O_2 and increased production of highly reactive oxygen species. Among these, hydroxyl radicals (HO•) and singlet molecular oxygen ($^1O_2$) are considered to be the most reactive and the most likely immediate oxidants of polyunsatured fatty acids side-chains of phospholipids.

In view of the hyperoxic state found in swimbladders (Wittenberg et al. 1980) in which the elevated oxygen pressure apparently does not harm the cells lining its walls, one would anticipate finding an extremely effective antioxidant mechanism operating in tissue exposed to elevated oxygen tension. Recent observations with human lung tissue have shown a high concentration of superoxide dismutase which increases upon exposure to high oxygen pressures (Huber and Draht 1981). Comparatively little information is known for the antioxidant defense mechanism in swimbladder or in the choroidal gland of the eye in fish which are known to be exposed to elevated oxygen pressures.

In this paper, we report some initial observations on the fatty acid composition, glutathione content and antioxidant activity in swimbladder homogenates.

2 Materials and Methods

Salmo trutta was kindly provided by Professor L. Bolis, University of Messina Biological Laboratories Messina, Italy. Swimbladder were removed from the fish on glass lined ice blocks, weighed and homogenized in 50 mM potassium phosphate buffer, pH 7.6.

1 Institute of Biological Chemistry, University of Catania, V. le A. Doria, 6 – 95125 Catania, Italy
2 Institute of Pharmaceutical, University of Catania, V. le A. Doria, 6 – 95125 Catania, Italy

Toxins, Drugs, and Pollutants in Marine Animals
(ed. by Bolis et al.)

2.1 Extraction of Lipids

Lipids were extracted from swimbladder homogenates by a modified procedure (Bligh and Dyer 1959). A solution of 50 ml chloroform: methanol (2:1) was added to the swimbladder homogenate (2 g). The suspension was shaken vigorously and the one phase suspension was allowed to stand overnight. Then 37.5 ml of chloroform and 37.5 ml of 1.0 M KCl solution containing 0.4% (v/v) glacial acetic acid was added and the mixture shaken. After several hours, the mixture separated into two phases, the lower phase containing the lipids was filtered through a Whatman no. 12 filter paper. Lipids were analyzed for phosphate as described by White and Frerman (1967).

2.2 Mild Alkaline Methanolysis

Mild alkaline methanolysis was used to cleave the diacyl phospholipids into water soluble glycerol phosphate esters and fatty acid methyl esters (Metcalfe and Schmitz 1961). The fatty acid methyl esters were recovered from the methanolysis mixture by three extractions with diethyl ether and final extraction with chloroform.

2.3 Gas Chromatography

A Carlo Erba Fractovap 4200 gas chromatography was used to separate and identify the various fatty acid methyl esters. Glass columns 3 mm O.D. × 1.8 m were packed with 3% SE–30 on 80–100 mesh CWHP. The elution profile of fatty acids was obtained by temperature gradient of $20^{o}C\ min^{-1}$ between 140° and $260^{o}C$.

2.4 Ultraviolet Absorption Spectra

All spectrophotometric measurements were performed on a Cary 219 spectrophotometer. Diene conjugation absorption of swimbladder total lipids as well as lipids extracted from the choroidal gland, gills and liver were measured at a lipid concentration of 1 mg ml^{-1} in n-hexane. $FeCl_3$-EDTA oxidized lipids were measured in a similar fashion. The scan rate was maintained at 20 nm min^{-1} and the bandwidth was 0.1.

2.5 Antioxidant activity

Oxidation of lipids assayed as malonaldehyde production was promoted in the presence of 50 $\mu g\ ml^{-1}$ xanthine oxidase, 0.35 mM hypoxanthine, 0.1 mM $FeCl_3$ and 0.1 mM EDTA-Fe. The reaction mixture contained 0.5–1 μmol of lipid phosphorus ml^{-1} and 0.25 M Tris-HCl buffer pH 6.8. The incubation time was 1 h at $37^{o}C$. Antioxidant activity in swimbladder homogenate was evaluated by the extend of inhibition of malonaldehyde production or as a lowering of the 233 nm absorption maximum of the conjugated diene spectrum.

Malonaldehyde was estimated with the thiobarbituric acid test (Uchiyama and Mihara 1978), after oxidizing lipids with $FeCl_3$–EDTA in the presence or absence of swimbladder homogenates. The reaction mixture for measuring malonaldehyde contained 0.5 ml of oxidized lipids to which was added 3 ml of 1% phosphoric acid and 1 ml of 0.6% thiobarbituric acid aqueous solution. The mixture was heated for 45 min in a boiling water bath. After cooling, 4 ml of n-butanol was added and the mixture vigorously shaken. The butanol phase was separated by centrifugation and absorbance was measured at 535 and 520 nm. The difference was used as the thiobarbituric acid value.

Reduced Glutathione Concentration was measured in fresh swimbladder homogenates by the method of Ellman et al. (1961). Homogenates were prepared in 5% TCA and after centrifugation the supernatant was extracted five time with diethyl ether and reduced glutathione was measured at 412 nm after reacting with DTNB.

Proteins were estimated by the method of Lowry et al. (1951). All reagents used were the purest grade available commercially.

3 Results

The fatty acid composition of swimbladders from *Salmo trutta* was compared to that of the choroidal gland, gills and liver. Table 1 illustrates the amount of unsaturated fatty acids in the various organs from *Salmo trutta.* The data are expressed as weight percent of total extractable phospholipid phosphate. As can be seen, the swimbladder and choroidal gland contains the lowest percent of unsaturated fatty acids (52%) as compared to gills (60%) and liver (70%).

Since it is generally recognized that unsaturated fatty acids are liable to peroxidation because of the presence of carbon-carbon double bonds which weaken the carbon-hydrogen bond on the adjacent carbon atom, studies were undertaken to determine if conjugated dienes could be detected in the lipid extracts prepared from the various organs of *Salmo trutta.* Figure 1 shows the spectra of the lipid extracts (dashed line). As can be seen, the lipids from the swimbladder and gills do not reveal the presence of conjugates dienes whereas, liver and choroidal gland show significant absorption in the 230–233 nm range suggestion that these lipids are more sensitive to the routine extraction procedures. However, when the lipid extracts were subjected to peroxidative oxidation with $FeCl_3$-EDTA (solid line, Fig. 1), a significant absorption peak at 233 nm was observed for the liver and gills whereas, it was less pronounced for the swimbladder lipids. The peroxidation of lipids from the choroidal gland was not performed.

These experiments clearly demonstrated that the 233 nm absorption peak is due to the presence of peroxidized lipid fractions which manifest conjugated diene structures.

The observation that swimbladder lipids are less easily peroxidized, as shown by the 233 nm absorption maximum, suggested that lipid soluble as well as water-soluble antioxidants might be involved in maintaining the structural integrity of swimbladder lipids. To this end, we have carried out a series of experiments with swim-

Table 1. Fatty acid composition in lipid extracts from different organs of *Salmo trutta*

Carbon no	14	16:1	16:0	18:1	18:0	20 unsat.	20:0	22 unsat.
Swimbladder	8.16	8.5	24.7	28.6	12.4	8.4	5.1	6.9
Choroidal gland	9.3	7.8	26.1	25.4	11.8	9.5	4.3	6.6
Gills	4.5	8.7	22.8	30.8	10.2	12.1	--	10.8
Liver	--	3.9	18.5	14.7	7.7	30.5	--	24.6

Data expressed as weight percent of total phospholipid phosphate

bladder homogenates incubated with total lipid extract from choroidal gland, gills or liver which were subjected to peroxidative oxidation in the presence of xanthine oxidase, hypoxanthine, and $FeCl_3$-EDTA. Table 2 illustrates the results of these studies. As can be seen, swimbladder homogenates added to the complete peroxidation system significantly decreases the amount of molonaldehyde produced. Furthermore, the inhibitation of malonaldehyde production was linear with the mg of the Lowry proteins ml^{-1} of swimbladder homogenate incubated in the reaction mixture. The inhibitory

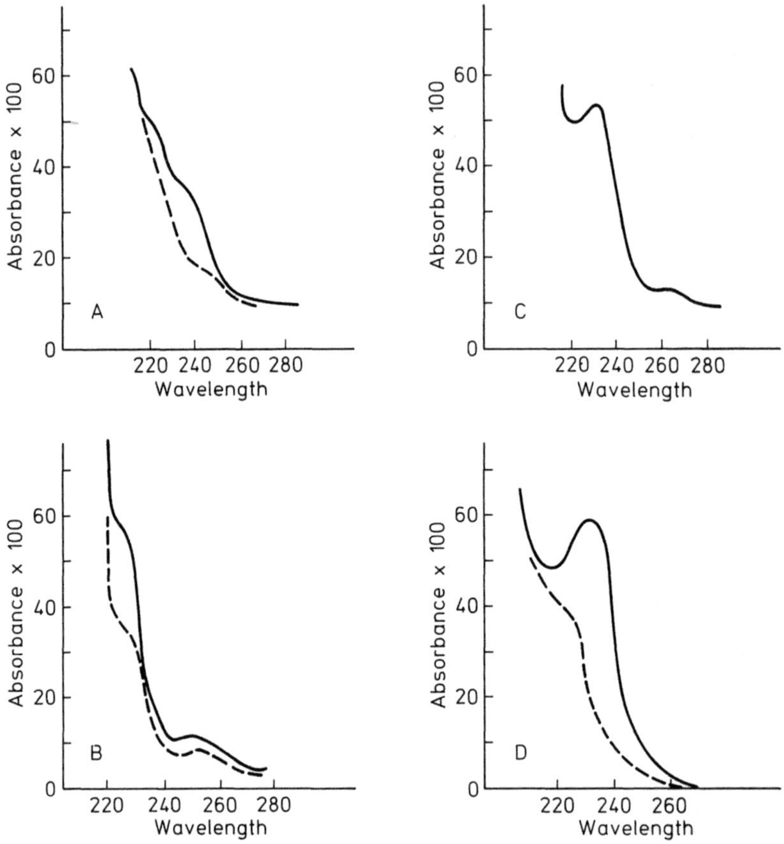

Fig. 1. Diene conjugation absorption. **A** Swimbladder; **B** gills; **C** choroidal gland; **D** liver. ---- Total lipid extract lmg ml^{-1} ; ——— total lipid extract subjected to Fe Cl_3 - EDTA peroxidation

Table 2. Malonaldehyde formation

Reaction system	nmol ml^{-1} min^{-1}
Complete	14.8
+ Swimbladder homogenate 50 μg ml^{-1} Lowry protein	3.1
+ 3 μg ml^{-1} SOD	4.4
– $FeCl_3$ EDTA	0.30
– Xanthine oxidase	0.93

Complete reaction system contains 0.35 mM hypoxanthine 50 μg ml^{-1} xanthine oxidase, 0.5 μmol of lipid phosphorous ml^{-1} of lipid extract from swimbladder gills, liver or choroidal gland, 0.1 mM $FeCl_3$ and 0.1 mM EDTA. The reaction was initiated by the addition of xanthine oxidase, and malonaldehyde. Production of malonaldehyde was determined as described in Methods.

Table 3. Reduced glutathione (μmol mg^{-1} wet weight of tissue)

Swimbladder	0.13 ± 0.02
Chroidal gland	0.08 ± 0.04
Gills	0.28 ± 0.06
Liver	0.46 ± 0.02

Results are expressed as mean ± s; d; for five separate fish

capacity of the swimbladder homogenate was destroyed upon heated at 80°C for 15 min. This observation suggested that the enzyme superoxide dismutase (SOD) might be present in the homogenate. The results of these experiments (Table 2) show that peroxidation activity promoted by xanthine oxidase was inhibited by swimbladder homogenates, comparable to the inhibition procedure with 3 μg ml^{-1} of SOD.

The inhibitory activity displayed by swimbladder homogenates in lipoperoxidation assays is currently being examined in order to establish if this inhibition is due to the presence of superoxide dismutase.

Reduced glutathione concentrations were measured in various organs of *Salmo trutta*. Table 3 shows that the choroidal gland and swimbladder contain the lowest concentrations of reduced glutathione as compared to gills and liver. This measurement at present does not allow one to conclude with certain that sulfhydryl oxidation is not without significant effect in hyperoxigination in vivo. Indeed, an increase in glutathione turnover as been shown in frog bladder preparations (Allen et al. 1973) and in the isolated lung preparations (Nishihi et al. 1976), immediately upon raising oxygen pressure. Further work is needed in this comparatively little known area of sulfhydryl chemistry correlated to hyperoxic states.

4 Discussion

The general theory of oxygen toxicity formulated by Gerschman (1954), proposed similarities between cellular alterations produced by radiation and free radical intermediates of molecular oxygen. The discovery of superoxide dismutase (Fridovich 1974) has given further strength to Gerschman's theory that aerobic organisms are equipped with enzymatic "scavengers" for the highly reactive superoxide free radical which are produced during reduction of molecular oxygen.

Our studies suggest that swimbladders contain antioxidant mechanisms which inhibit peroxidative oxidation of polyunsaturated fatty acids. The finding that swimbladder homogenates contain a heat-labile, water-soluble antioxidant would suggest the presence of a superoxide dismutase enzyme.

The protective role of sulfhydryl compounds and other agents against cellular damage due to radiation or superoxide anion production has been previously described (Barton 1955). Significant reductions in SH/S-S ratios have been demonstrated in rat lung tissue after exposure to 5 atm of oxygen for 45 min (Jamieson et al. 1963). Similarly, there is ample evidence that high pressure oxygen inactivates sulfhydryl enzymes in various in vitro preparations. Our findings that glutathione levels were lower in swimbladder than gills or liver might suggest an increased turnover of glutathione at elevated oxygen pressure as observed for the frog bladder (Allen et al. 1973). Furthermore, the presence of elevated activities of hexose monophosphate shunt enzymes in the swimbladder would suggest an elevated production of NADPH for maintaining essential-SH groups in the reduced state. Further work in progress for estimating glutathione turnover tates in swimbladder and choroidal gland.

References

Allen JE, Goodman DEP, Besarab A, Rasmussen H (1973) Studies on biochemiochemical basis of oxygen toxicity. Biochim Biophys Acta 320: 708–728

Barton ESG (1955) Oxidation of some oxidation – reduction systems by oxygen at high pressures. Arch Biochem Biophys 59: 502–510

Bligh EG, Dyer WJ (1959) Analysis and extraction of tissue phospholipids. Can J Biochem 37: 911–916

Ellman GL, Courtney DK, Anders V Jr and Featherstone PM (1961) A nex and rapid colorimetric determination of acetylcholinesterase activity. Biochem Pharmacol 1: 88–91

Fridovich I (1974) Superoxide dismutase. Adv. Enzymol Relat Areas Mol Biol 41: 35–87

Gerschman R, Gilbert DL, Nyl SW, Dwyer P, and Fenn WO (1954) Oxygen poisoning and x-irradiation: a mechanism in common. Science (Wash DC). 119: 623–626

Gilbert D (1981) Oxygen: an overall biological view in oxygen and living processes. Gilbert L Springer, Berlin Heidelberg New York, pp 376–392

Haldane JS, Priestly JC (1935) Respiration 2nd edn. Conn. Yale University Press, New Haven

Huber G, Draht DB (1981) Pulmonary oxygen toxicity in oxygen and living processes. Gilbert DL (ed). Springer, Berlin Heidelberg New York , pp 273–342

Jaimeson D, Ladner K, Van Den Brenk HAS (1963) Pulmonary damage due to high pressure oxygen breathing in rats IV. Quantitative analysis of sulfhydryl and disulfide groups in rat lungs. Aust J Exp Biol Med Sci 41: 491–497

Lowry OH, Rosebrough NJ, Farr AL, Rose J, Randall M (1951) Protein measurements with folin phenol reagent. J Biol Chem 193: 265–275

Metcalfe LD, Schmitz AA (1961) The rapid preparation of fatty acid esters for gas chromatographic analysis. Anal Chem 33: 363–369

Nishiki K, Jamieson D, Oshino N, Chance B (1976) Oxygen toxicity in perfused rat liver and lung under hyperbaric conditions. Biochem J 160: 343–345

Uchiyama M, Mihara M (1978) Determination of malonaldehyde precursor in tissues by thiobarbituric acid test. Anal Biochem 86: 271–278

White DC, Frerman FE (1967) Phospholipid composition in *H. parainfluenzal* J'Bacteriol 94: 1854–1861

Wittenberg JB, Copeland DE, Haedrich RL, Child JS (1980) The swimbladder of deep sea fish: the swimbladder wall is a lipid-rich barrier to oxygen diffusion. J Mar Biol Assoc UK 60: 263–276

Effects of Metal Ions on Respiratory Structures

K. SIMKISS[1]

1 Introduction

Respiratory surfaces have evolved so as to present an extensive surface area to a large volume of environmental gases. For aquatic organisms this is usually achieved by an elaborate epithelium separating, by a short diffusion path, a counter current flow of blood and water. The properties of such surfaces have led to them being variously adapted for additional functions such as the uptake of dissolved metabolites (Stephens 1982), the collection of suspended particles (Jorgensen 1955) or the regulation of electrolyte balance (Evans 1982). The mechanisms whereby these multiple functions have been achieved have been the subject of extensive research. What have received less attention are the possible mechanisms whereby organisms protect themselves from the consequences of the inadvertent uptake of pollutant materials brought onto this epithelium by the respiratory currents.

This article will ruthlessly pursue a single line of thought, not because it is necessarily entirely correct but because the balance of research is so heavily in the other direction, that this aspect needs to be emphasized. The concept that will be pursued is that whereas the metals of most interest to physiologists have been the strongly ionic "class a" type (e.g., Na, K, Ca, Mg) those of most interest to enzymologists and pollution biologists are the strongly covalent "borderline or class b" metals (e.g., Cu, Zn, Cd, Hg). The differences for biology between these two types of metal are enormous. The "class a" metals are the bulk electrolytes and exist at concentrations many orders of magnitude above those of the "class b" metals. The biological activities of the "class a" metals are dictated by membrane phenomena whilst those of "class b" are much more involved in ligand binding.

Let us start by considering the normal model for metal ion/plasma membrane interactions (Fig. 1). The ions in sea water are considered to be derived from the salts of "class a" or the hard acid metals (Na, K, Ca, Mg) which dissociate in water to form highly charged and hydrated particles. The cell membrane is basically a lipid bilayer with a charged surface on which a number of layers of counter-charged ions are adsorbed. In order for a metal ion to enter the cell it must overcome first the electrical field over the surface and then penetrate the lipid bilayer. Since metal ions are usually considered to be heavily charged hydrophilic particles and membranes are hydro-

1 Department of Pure and Applied Zoology, University of Reading, Whiteknights, Reading RG6 2 AJ England

Toxins, Drugs, and Pollutants in Marine Animals
(ed. by Bolis et al.)

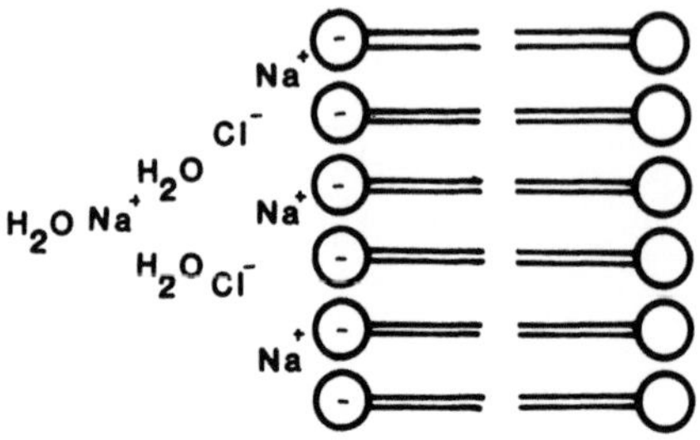

Fig. 1. The movement of sodium ions across a lipid bilayer is impeded by the ions adsorbed onto the surface of the membrane and by the hydration shell around the cation

phobic sheets there is an enormous energy barrier to the movement of ions from the medium into the cell. As a result membranes can be regarded as barriers separating large sources of electro-chemical energy. Many of the properties of living systems are dependent upon the controlled release of these energy dependent gradients to initiate ATP synthesis, muscular contraction, neural activity, electrolyte balance, etc. Clearly what we are describing here are the fundamental properties of cell physiology but what we should consider now is whether they are the ones of greatest importance to respiratory surfaces.

2 Ion Pairs in Sea Water

The inorganic environment of the cell includes not only fully dissociated ions but also ion pairs such as $NaCl^o$, $CaSO_4{}^o$ and $NaHCO_3{}^o$. These consist of an anion and a cation temporarily coalescing through mutual electrostatic attraction so as to lose some of their charged nature and some of their water of hydration. As such they would be expected to show abnormal membrane permeabilities and this does in fact appear to be the case (Burton 1983). When one considers either the essential trace metals (e.g., Cu, Zn, Co) or the worst pollutants (e.g., Hg, Cd) the phenomenon of ion pair formation becomes progessively more important. The "class b" or soft acid metals are characterised by having a progessively increasing covalent basis to their chemistry. As a result of this, sea water or physiological salines containing trace amounts of these metals, consist of metal species such as $(MCl)^+$, $(MCO_3)^o$, $(MSO_4)^o$, $(MCl_2)^o$, $(MClOH)^o$ and $(MCl_3)^-$. There are considerable problems in analysing these complexes, but a number of attempts have been made to determine them and, as might be expected, they are dependent upon the pH and anion concentrations present (Zirino and Yamamoto 1972; Whitfield and Turner 1981). Under normal physiological circumstances, however, one could summarise the evidence by saying that the bulk of these trace and pollutant metals do not normally exist as ions but as inorganic covalent complexes. It is quite likely, therefore, that the normal concepts of ion physiology, as summarised in Fig. 1, do not apply.

3 Membrane Penetration of Ion Pairs

The penetration of ion pairs through membranes has been considered intermittently as an aspect of electrolyte balance. From what has just been argued, however, it may be a very important aspect of the uptake of trace and pollutant metals. In order to emphasize this let us consider an extreme form, namely the penetration of mercury through cell membranes. Mercury is a particularly interesting metal since it has a co-ordination number of 2. Under physiological conditions the most common metal species is $(HgCl_2)^o$ which has lost both its charge and its coordinated water molecules. In discussing the penetration of $(HgCl_2)^o$ through membranes I will depend largely on the elegant experiments of Gutknecht (1981). In this work artificial "black" membranes were used and shown to have remarkably high fluxes for ^{203}Hg. The fluxes were analysed on the basis that a membrane and its associated unstirred layers were analogous to conductance in series

i.e.,
$$\frac{1}{J_{Hg}} = \frac{1}{p^{ul}\,Hg^{t}} + \frac{1}{p^{m}\,HgCl_2}$$

or
$$\frac{1}{p^{t}} = \frac{Hg^{t}}{p^{m}\,HgCl_2} + \frac{1}{p^{ul}}$$

where
Hg^t = total concentration of mercury
$HgCl_2$ = concentration of $HgCl_2$
J_{Hg} = total mercury flux
p^{ul} = unstirred layer permeability coefficient
p^m = membrane permeability to $HgCl_2$
p^t = total mercury permeability coefficient

Normal values of p^m for electrolytes are about 10^{-11} cm s^{-1} but in a saline of 100 mmol NaCl dm^{-3} with a mercury concentration of 100 μmol dm^{-3} Gutknecht found that p^t was about 10^{-4} cm s^{-1} (which is similar to that expected through an aqueous unstirred layer of about 100 μm) while p^m was found to be 1.3×10^{-2} cm s^{-1}.

The important conclusion therefore is that $(HgCl_2)^o$ passes across membranes more than a million times faster than ions such as Na^+. Its movement is, however, controlled by the aqueous unstirred layer which dictates the rate at which $(HgCl_2)^o$ is delivered to the membrane surface, Thus any factor that affects the concentration of $(HgCl_2)^o$ on this surface will instantly affect the flux. The total mercury concentration is important in this context, but so also is the ability to transform any other metal species that traverse the unstirred layer into the permeant form. This includes therefore the occurrence of chloride ions and protons on the membrane surface.

The high permeability of cell membranes to $(HgCl_2)^o$ effectively means that they cease to function as compartmental barriers to this molecule and the cells are effectively "transparent" to these metal species. The control of the entry of mercury into cells will therefore depend mainly upon the properties of the unstirred layer on the gill epithelium and the probability of other mercury compounds being converted into the permeant $(HgCl_2)^o$ form at the surface of the cell. This means that close attention should in future be paid to the microanatomy of the gill in terms of (a) mucus layers (providing an unstirred aqueous barrier to pollutants) and (b) the site of "chloride cells" or other ion transporting systems which may affect the metal speciation by affecting the charge ratio

$$\text{i.e.,}\quad MCl^+ \rightleftharpoons \underset{\text{permeant}}{(MCl_2)^o} \rightleftharpoons MCl_3^-$$

An epithelial cell that is transporting chloride ions beneath the mucus layer could theoretically impede the entry of mercury in both marine and freshwater fish by driving the reaction from $(HgCl_2)^o$ into one or other of these charged forms. It is therefore interesting to note that there is some evidence that fish mucus can maintain such gradients (Shephard 1982) that mucus secretion may increase in metal polluted fish (Eddy and Fraser 1982) and that both copper and zinc ions have been shown to induce chloride cell populations on the secondary lamellae of fish gills (Baker 1969; Matthiessen and Brafield 1973).

It should be remembered, of course, that these discussions have centred upon mercury. Other pollutant metals have coordination numbers greater than 2 and thus retain some hydration shells as well as having different relationships with a variety of anions. It may, therefore, be worth considering another example based upon quite different data.

The common mussel, *Mytilus edulis*, is capable of accumulating high concentrations of cadmium from the surrounding sea water. Using the isolated gill, Carpene and George (1981) studied the initial uptake of 10 μmol ^{109}Cd dm^{-3} in sea water. Uptake was proportional to external concentrations, showed no energy dependence or temperature effects and was independent of other ions. They concluded, therefore, that entry into the gill was by passive diffusion of metal species such as $(CdCl_2)^o$ and they calculated a diffusion constant of about 10^{-6} cm s^{-1} which is similar to that for the diffusion of cadmium in sea water.

One other feature should be noted before leaving the concept of metal ion pairs, and that is that we are dealing with a complicated chemical situation. We have recently shown that most of the trace metals in sea water are lipid-soluble (Simkiss 1983). If such solutions are shaken with olive oil the metals appear to show an oil/water partition in excess of unity. The situation in these experiments is undoubtedly complex since in order to show this effect the lipids have to be "donor solvents" and the molecules in these experiments are, of course, free to rotate in three dimensions unlike the planar lipids of the cell membrane. The experiments are interesting, however, because they not only emphasize this unusual property of lipid solubility but also demonstrate that we are not just dealing with the thermodynamics of an oil/water partition but also with the kinetics of the formation of permeant molecules. The ion pairs are continually being removed into the lipid layer but the rate at which this occurs depends upon the kinetics that determine the rate at which they are re-formed from other ions in solution. The system never fully equilibrates and the same is true, of course, for gills. The respiratory current ensures that any permeant metal species are continually being made available to the membranes of the epithelium but whether they are replaced during the time of a respiratory cycle is a kinetic problem. It is interesting to note, however, that zinc uptake may parallel oxygen uptake over a considerable range of metabolic rates (Matthiessen and Brafield 1977).

4 Acid Effects

Protons can, of course, be considered as metal ions. In fact the same arguments that have been applied to the penetration of metals into cells are applicable to protons. The permeability coefficient of protons through artificial membranes is roughly 10^{-9} cm s^{-1}. Nitric acid in the form $(HNO_3)^0$ has a permeability of 10^{-4} cm s^{-1} and hydrochloric acid in the form $(HCl)^0$ is extremely permeable with a coefficient of 3 cm s^{-1} (Gutknecht and Walker 1981). One is, therefore, considering undissociated acid molecules as again being millions of times more permeant than Na^+, K^+ and similar ions. Clearly the question that arises is the availability of these uncharged complexes in environmental situations and certainly on this basis nitric acid should be more toxic in "acid rain" than sulphuric acid; the other main component. This assumes, that the toxicity of acids to cells depends upon their penetration into the cytoplasm. Protons are, however, capable of displacing other metals from a variety of ligands and there is a considerable amount of evidence that epithelia such as gills become leaky because calcium mediated intercellular adhesion is reduced. (McWilliams and Potts 1978; McWilliams 1980).

5 Cytoplasmic Responses to Metals

If it is accepted that there is some doubt about the effectiveness of the plasma membrane in regulating the entry of many trace metals into the cell then the emphasis for the control of trace metal metabolism shifts from the membrane to the chemistry of the protein ligands in the cell. Again it should be emphasized that whereas "class a" metals dictate the electro chemical and osmotic functions of the cell, "class b" metals are involved in many of the normal enzyme functions. Class b metals are more reactive to sulphur and nitrogen ligands than those of "class a" and they are therefore extremely important not only in normal metalloenzyme function but also in poisoning cell metabolism. If one envisages a metalloprotein in an "entatic" state then substituting the wrong metal or deforming the protein by the secondary binding of other metals (Fig. 2) will effectively destroy its catalytic abilities. One must therefore consider what happens to the respiratory epithelium if a variety of "class b" metals become applied to it.

The enzymes which have attracted greatest attention in the fish gill are carbonic anhydrase and the transport ATPases. Carbonic anhydrase is a zinc containing en-

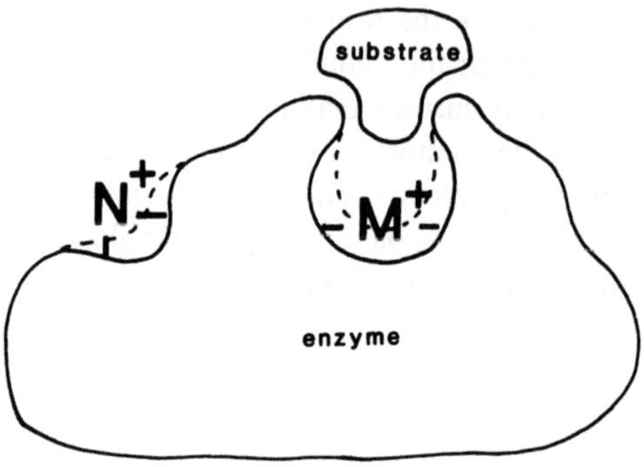

Fig. 2. The activity of a metalloenzyme may be affected by the substitution of other metals at the active site *M* or by the binding of metals onto the surface *(N)*. In both cases there is distortion of the enzyme which reacts less favourably with the substrate

zyme which catalyses the hydration of carbon dioxide to carbonic acid. It provides an excellent example of the importance of specific metals attaching to particular ligands, for in a series of in vitro experiments, Coleman (1967) replaced the zinc in the enzyme with a series of other metals. The activity of the resulting enzymes were 56% for cobalt and less than 5% for nickel, cadmium, manganese and copper substituted molecules. Clearly if the wrong metal is able to become incorporated into site M on a metalloenzyme (Fig. 2) its functioning becomes distorted.

The other effect of "class b" metals on enzymes can be illustrated by the fact that a number of membrane transport proteins may still function when exposed to pollutant metals. Thus in the crab *Carcinus maenas* cadmium may be transported into the gill on the calcium pump (Wright 1977). In the roach *Rutilus rutilus* Shephard and Simkiss (1978) found that Cu^{2+}, Zn^{2+} and Hg^{2+} completely inhibited the activity of Ca^{2+} ATPase in vitro but in vivo they actually enhanced the number of enzyme units available. Thus exposing fish to 0.157 μmol Cu^{2+} dm^{-3} for a period of 56 days increased the protein in the enzyme fraction of the gills by 78% and increased the potential enzyme activity by about 40%. A similar increase in Na/K ATPase occurred following exposure of *Platichthys flesus* to Cu^{2+} in vivo (Stagg and Shuttleworth 1982). These experiment can be interpreted as showing that whereas pollutant metals may enter cells by parasitizing active transport systems they also denature some of the proteins (Fig. 2) with the result that additional enzyme units have to be synthesized in order to maintain gill function (Table 1).

A second interesting interaction also occurs in a large number of respiratory epithelia. In the presence of metals such as Cu^{2+}, Zn^{2+}. Hg^{2+} and Cd^{2+} there is enhanced synthesis of metallothionein (Kito et al 1982a) or similar metal binding proteins (Roesijadi 1981, 1982). Proteins of this type are rich in sulphydryl groups and avidly bind large quantities of these metals. Opinions differ as to whether these proteins function as storage molecules for the essential metals Cu^{2+} and Zn^{2+} or whether they are detoxification systems for removing interfering metals such as Cd^{2+} and Hg^{2+} (Simkiss et al. 1982). It has been shown on a number of occasions however that fish exposed to sub lethal levels of Cd^{2+}, Hg^{2+} or Zn^{2+} develop some tolerance to subsequent doses (Duncan and Klaverkamp 1983) and this has been attributed to the induction of metallothionein synthesis (Kito et al. 1982b). We have recently shown that the gills of the mollusc *Littorina littorea* contain increasing amounts of

Table 1. Effects of copper on metal ATPase activities of fish gills. Values have been normalized to percentages to simplifiy interpretation. (Data from Shephard and Simkiss 1978, Stagg and Shuttleworth 1982)

	Controls	Copper treated	t test	Implication
NA/K ATPase (activity/unit protein)	100%	118%	n.s.	Fish maintains gill activity
Na/K ATPase (ovabain binding sites)	100%	151%	***	Increased numbers enzyme sites
Ca ATPase (activity/unit protein)	100%	81%	n.s.	Fish maintains gill activity
Ca ATPase (activity/filament)	100%	139%	***	Increased numbers enzyme sites

copper and zinc if they come from sites polluted with these metals. Ultrastructural studies of the cells of these ctenidia show copper deposits associated with sulphur in lysosomal bodies (Mason et al. 1984). Such observations are not uncommon and appear to result from the rapid turnover of metallothioneins. These proteins are typical of many cytoplasmic proteins in having a half life of only a few days. They are degraded by lysosomes and resynthesized within the cell but the metals associated with these proteins often accumulate within the residual bodies. This raises two interesting alternatives. Are these specific proteins present in all respiratory structures "blocking" the penetration of metals and thus preventing their entry into the vascular system? Alternatively are the protective systems in gills relatively poor so that cytoplasmic proteins are continually being denatured by "class b" metals? In this case the turnover of cytoplasmic proteins in the gill should be particuarly rapid. Either of these interpretations presents a view of gill physiology that is so different from the usual model of a respiratory/ion regulatory system that it is worth summarizing some of these features.

6 Conclusions

Figure 3 illustrates some of the cellular concepts that have been discussed in relation to fish gills. On the outer surface is a layer of mucus which has the effect of converting gradients that are maintained by hydraulic currents into fluxes that are controlled by diffusion pathways. It is the first and most obvious sign that the gills are not solely involved in respiration for while it may have a number of functions (Hughes 1980) it frequently acts as an inhibitor of respiratory gas exchange. This is because it provides an unstirred layer that might then reduce the extraction of lipid soluble molecules into the plasma membranes of the gills. The extent to which this succeeds in the case of pollutant metals will depend upon:

a) The effects that the metals have in interacting and binding to the ligand components of the mucus;
b) The extent of any ion gradients that exist across the mucus layer; and,
c) The kinetics of the formation of permeant species on both sides of the unstirred layer.

Those metal species that reach the gill epithelium will affect both the intercellular and the transcellular routes of ion movement. Antagonists of calcium will greatly affect the intercellular route and this is clearly seen with protons, zinc, mercury and perhaps aluminium (Muniz and Leivestad 1980) all of which have been shown to make the epithelium leaky and all of which can be partly antagonized by elevated levels of environmental calcium ions. The alternative transcellular route of metal uptake will not only allow pollutants to reach the vascular system but will also potentially inhibit the normal cellular activities of the gill itself.

It is perhaps surprising that very little appears to be known about the cellular dynamics of the gill epithelium. This is in contrast for example, to the intestinal mucosa which has a similar exposure to environmental compounds. The cells lining the ali-

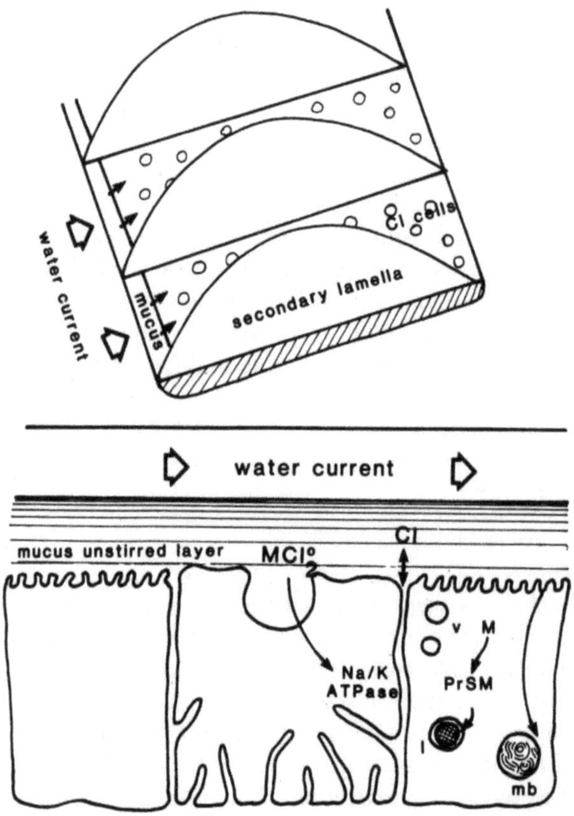

Fig. 3. *Top* the respiratory water current passes over the primary lamella, where mucus is secreted over the epithelial and chloride cells, and between the secondary lamellae. (After Hughes 1980). *Bottom* the mucus layer acts as an unstirred layer impeding the diffusion of metal complexes (MCl_2^o) into the cells. The availability of chloride complexes may be affected by the activity of the chloride cells. Metal complexes that enter the cells may inhibit transport ATPases thereby reducing ion and osmo-regulation and inducing cell vacuolation *(v)*. Metals *(M)* may be detoxified by metallothionein-like proteins which accumulate in lysosomes *(l)*. There may also be increased turnover of the cell membranes and the accumulation of fragments in myelin bodies *(mb)*

mentary tract have a turnover time of a few days, but there appears to be very little comparable information for the gill epithelium. Most mammalian cells also have enzymes with half lives measured in minutes to a few days. The situation in the gill epithelium is again unknown. What is clear, however, is that in the presence of metal pollutants:

a) The synthesis of transport ATPase's is increased (Table 1). The Ca ATPase enzyme can be stimulated by zinc ions (Watson and Beamish 1981) and it has been suggested that ATPases in chloride cells can also excrete some ions (Matthiessen and Brafield 1977)
b) Metals accumulate in the cytoplasm to levels that are many times greater than environmental concentrations
c) Metallothioneins are synthesized and pollutant metal fluxes reduced
d) There is considerable recycling of cytoplasmic proteins via the lysosomal system which accumulates metals
e) There is evidence that the cells own volume regulatory system is impaired (i.e, vacuolation occurs) and there is increased turnover of membrane systems (myelin bodies are found)
f) There is a switch in cellular differentiation and chloride cells are found on secondary lamellae.

All the indications are, therefore, that the gills show a very dynamic response to metal pollution. It must be admitted, however, that the information available to support these interpretations and conclusions is sparse, but worthy of further study.

References

Baker JTP (1969) Histological and electron microscopical observations on copper poisoning in the winter flounder *(Pseudopleuronectes americanus)*. J Fish Res Board Can 26: 2785–2793

Burton RF (1983) Inorganic ion pairs in physiology: significance and quantitation. Comp Biochem Physiol 74A: 781–786

Carpene E, George SG (1981) Absorptiom of cadmium by gills of *Mytilus edulis*. Mol Physiol 1: 23–34

Coleman JE (1967) Metal ion dependent binding of sulphonamide to carbonic anhydrase. Nature (Lond) 214: 193–194

Duncan DA, Klaverkamp JF (1983) Tolerance and resistance to cadmium in white suckers *(Catostomus commersoni)* previously exposed to cadmium, mercury, zinc or selenium. Can J Fish Aquatic Sci 40: 128–138

Eddy FR, Fraser JE (1982) Sialic acid and mucus production in rainbow trout *(Salmo gairdneri)* in response to zinc in sea water. Comp Biochem Physiol 73C: 357–359

Evans DH (1982) Salt and water exchange across vertebrate gills. In: Houlihan DF, Rankin JC and Shuttleworth TJ (eds) Gills. Camb Univ Press

Gutknecht J (1981) Inorganic mercury (Hg^{2+}) transport through lipid bilayer membranes. J Membr Biol 61: 61–66

Gutknecht J, Walter A (1981) Transport of protons and hydrochloric acid through lipid bilayer membranes. Biochim Biophys Acta 641: 183–188

Hughes GM (1980) Functional morphology of fish gills. In: Lahlou B (ed) Epithelial transport in the lower vertebrates. Camb Univ Press pp 15–36

Jorgensen CB (1955) Quantitative aspects of filter feeding in invertebrates. Biol Rev Camp Philos Soc 30: 391–454

Kito H, Tazawa T, Ose Y, Sato T, Ishikawa T (1982a) Formation of metallothionein in fish. Comp Biochem Physiol 73C: 129–134

Kito H, Tazawa T, Ose Y, Sato T, Ishikawa T (1982b) Protection by metallothionein against cadmium toxicity. Comp Biochem Physiol 73C: 135–139

Mason AZ, Simkiss K, Ryan K (1984) Ultrastructural localization of metals in specimens of *Littorina littorea* (L.) collected from clean and polluted sites. J Mar Biol Assoc UK 64, 699–720

Matthiessen P, Brafield AE (1973) The effects of dissolved zinc on the gills of the stickleback *Gasterosteus aculeatus* (L.) J Fish Biol 5: 607–613

Matthiessen P, Brafield AE (1977) Uptake and loss of dissolved zinc by the stickleback *Gasterosteus aculeatus* (L.) J Fish Biol 10: 399–410

McWilliams PG (1980) Effects of pH and calcium on sodium uptake in Norwegian brown trout *(Salmo trutta)* from an acid river. J Exp Biol 88: 259–267

McWilliams PG, Potts WT (1978) The effects of pH and calcium concentrations on gill potentials in the brown trout *(Salmo trutta)* J Comp Physiol 126: 277–286

Muniz IP, Leivestad H (1980) Toxic effects of aluminium on the brown trout *Salmo trutta.* In: Drabløs D and Tollan A (eds) Ecological impact of acid rain precipitation. SNSF project, Norway, pp 320–321

Roesijadi G (1981) The significance of low molecular weight metallothionein-like proteins in marine invertebrates: current status. Mar Environ Res 4: 167–179

Roesijadi G (1982) Uptake and incorporating of mercury into mercury-binding proteins of gills of *Mytilus edulis* as a function of time. Mar Biol (NY) 66: 151–157

Shephard KL (1982) The influence of mucus on the diffusion of ions across the esophagus of fish. Physiol Zool 55: 23–34

Shephard KL, Simkiss K (1978) The effects of heavy metals on Ca^{2+} ATPase extracted from fish gills. Comp Biochem Physiol 61B: 69–72

Simkiss K (1983) Lipid solubility of heavy metals in saline solutions. J Mar Biol Assoc UK 63: 1–7

Simkiss K, Taylor MG, Mason AZ (1982) Metal detoxification and bioaccumulation in molluscs. Mar Biol Lett 3: 187–201

Stagg RM, Shuttleworth TJ (1982) The effects of copper on ionic regulation by the gills of sea water adapted flounder *(Platichthys flesus).* J Comp Physiol 149: 83–90

Stephens GC (1982) Recent progress in the study of "Die Ernährung der Wassertiere und der Stoffhaushalt der Gewässer". Am Zool 22: 611–619

Watson TA, Beamish WH (1981) The effects of zinc on branchial adenosine triphosphatase enzymes in vitro from rainbow trout *Salmo gairdneri.* Comp Biochem Physiol 68C: 167–173

Whitfield F, Turner D (1981) Sea water as an electrolyte medium. In: Whitfield J, Jagner D (eds) Marine electrochemistry. Wiley, New York

Wright DA (1977) The effect of calcium on cadmium uptake by the shore crab *Carcinus maenas.* J Exp Biol 67: 163–173

Zirino A, Tamamoto S (1972) A pH dependent model for the chemical speciation of copper, zinc, cadmium and lead in sea water. Limnol Oceanogr 17: 661–671

Heavy Metal Storage in Marine Animals Under Various Environmental Conditions

J.M. BOUQUEGNEAU[1], M. MARTOJA[2], and M. TRUCHET[3]

High loads of heavy metals can be found in some marine species either due to their peculiar physiology or due to the level of pollution of the medium they are living in. The first case is illustrated by copper loads of species using haemocyanin as oxygen-carrier and iron loads of species using haemoglobin. The other case is, for example, that of mollusc species living in the Severn Estuary, highly polluted by heavy metals: they contain more Cd, Cu and Zn than molluscs of the same species living in unpolluted mediums (Noël-Lambot et al. 1978).

Metabolic and toxic effects of metal ions have been reviewed in mollusca by Simkiss and Mason (1983). Excesses of essential metals as well as the presence in large amounts of non-essential ones, due to inappropriate intracellular binding, may be toxic for those species when the excess of metal is not stored in the animal under a detoxified form. Heavy metals can be stored an detoxified by marine organisms either by a compartmentation process within membrane-limited vesicles or by binding to specific proteins.

1 Storage of Metals by Binding to Specific Proteins

Since the work by Margoshes and Vallee (1957) about the intracellular cadmium binding in equine kidney, metallothioneins have been the most studies among metal binding proteins. Metallothioneins are soluble and heat-stable proteins. Their molecular weight, estimated by gel chromatography is about 10,000 but, estimated by the amino acid analyses, the molecular weight of the metal-free thionein is about 6,800 (Kojima and Kagi 1978; Roesijadi 1981).

The amino acid composition of metallothioneins reveals an unusually high cysteine content (till 30% of the total residues) and the absence of aromatic amino acids and histidine (Table 1). Metallothioneins are also characterized by a high metal content. Metals are probably bound to the thionein through mercaptide bonds, seven g-atoms metal being bound to one mole of protein (Sadler et al. 1978).

1 Chercheur Qualifié au Fonds National de la Recherche Scientifique. Laboratoire d'Océanologie, Université de Liege, B 4000 – Liege, Belgium
2 Institut Océanographique, rue Saint-Jacques, 195, 75005 – Paris, France
3 Laboratoire d'Histophysiologie fondamentale et appliquée, rue Cuvier, 12, 75005 – Paris, France

Toxins, Drugs, and Pollutants in Marine Animals
(ed. by Bolis et al.)

Table 1. Amino acid composition of some metal-binding proteins

	Rat liver	*Mytilus gallo-provincialis*	*Patella vulgata*		*Cancer pagurus*	*Scylla serrata*		Oyster	Rat liver 1
	Cu-thionein (1)	Cu-thionein (2)	MTa (3)	MTb	MT II (4)	MT I (5)	MT 2	Cd-protein (6)	Cu-chelatin (7)
1/2 Cys	30.0	17.4	21.0	20.0	29.9	26.2	28.3	7.6	7.5
Asp.	8.5	9.6	11.2	12.0	9.6	6.1	11.4	12.9	10.5
Thr.	3.8	7.7	7.2	6.8	7.3	5.9	4.8	5.8	7.3
Ser.	15.0	9.7	8.8	8.5	9.1	12.0	7.9	7.3	7.2
Glu.	6.3	5.5	7.7	8.5	5.6	10.9	9.7	9.7	11.3
Pro	0.0	4.7	3.8	4.1	8.5	7.2	11.3	4.4	2.5
Gly	7.3	18.2	11.0	10.6	7.3	9.4	6.4	9.5	9.4
Ala.	8.1	8.4	8.7	8.9	4.0	4.1	2.5	6.6	3.4
Val.	2.3	4.0	3.2	2.7	tr	1.8	0.3	6.1	6.8
Met.	1.8	0.0	<0.5	<0.5	1.3	0.0	0.2	1.5	1.7
Ile.	1.7	2.9	1.8	1.5				3.9	4.9
Leu.	0.7	1.5	2.6	2.7	0.0	0.6	0.2	4.5	4.0
Tyr.	0.0	0.0	1.8	1.8				2.6	1.7
Phe.	0.3	0.0	1.0	1.0				3.5	3.4
His.	0.0	1.3	1.0	1.4				8.9	15.9
Lys.	14.0	6.2	8.8	8.6	17.4	13.8	13.0	1.2	1.3
Arg.	0.0	2.9	0.8	0.9	0.0	1.8	3.7	2.9	1.3

1. Winge et al. (1981)
2. Viarengo ėt al. (1983)
3. Noël-Lambot et al. (1980)
4. Overnell and Trewhella (1979)
5. Olafson et al. (1979)
6. Ridlington and Fowler (1979)
7. Irons and Smith (1977)

Metals which bind to metallothioneins are bismuth, cadmium, copper, gold, mercury, silver and zinc, some of them being able to induce the synthesis of the protein.

Metallothioneins are widely distributed among prokaryotes and the animal kingdom (Roesijadi 1981) and can be found in some tissues of uncontaminated animals. Their physiological function is yet to be conclusively identified. However, it is generally accepted that this protein may play an important role in the intracellular regulation of the essential trace metals zinc and copper. Moreover, the ability of metallothioneins to be induced by, and bind large amounts of highly toxic metals has led to the hypothesis that this protein is involved in the sequestration and detoxification of such substances (Kojima and Kagi 1978; Roesijadi 1981).

To date, metalloproteins and particularly metallothionein-like proteins have been reported in several marine vertebrates (teleost fish: Olafson and Thompson 1974; Bouquegneau et al. 1975; Overnell and Coombs 1979; elasmobranchs: Flos and Hidalgo 1983) and invertebrates including Mollusca, Crustacea and Polychaetes (*Mytilus edulis*: Noël-Lambot 1976; George et al. 1979; Marshall and Talbott 1979; Frankenne et al. 1980; Roesijadi and Hall 1981; Viarengo et al. 1980; *Crassostrea virginica*: Casterline and Yip 1975; Ridlington and Fowler 1979; *Patella vulgata:* Howard and Nickles 1977; Noël-Lambot et al. 1978; Noël-Lambot et al. 1980; *Patella caerulea*: Noël-Lambot 1980; *Littorina littorea* and *Purpura lapillus*: Noël-Lambot et al. 1978; *Murex trunculus*: Bouquegneau et al. 1983; *Carcinus maenas*: Jennings et al. 1979; *Cancer pagurus*: Overnell and Trewhella 1979; *Cancer magister* and *Scylla serrata*: Olafson et al. 1979a; *Eudistylia vancouveri*: Young and Roesijadi 1983; etc).

However, to date, only some of those metallothionein-like proteins have been proved to be true metallothioneins. This has been, however, demontrated in the mollusca *Patella vulgata* and *Mytilus galloprovincialis* respectively by Noël-Lambot et al. (1980) and Viarengo et al. (1983) and in the crustacea *Scylla serrata* and *Cancer pagurus* respectively by Olafson et al. (1979a) and Overnell and Trewhella (1979) (see Table 1).

On the other hand, the low molecular weight cadmium binding protein of the oyster *Crassostrea virginica* described by Ridlington and Fowler (1979) is low in cysteine (8%), high in aspartate and glutamate and contains the aromatic amino acids and obviously is not characteristic of metallothionein (Table 1). Similarly, Lyon et al. (1983) have described metal binding proteins in the hepatopancreas of the crayfish (*Austropotamobius pallipes*) which are not typical metallothioneins.

Note that in the insects *Pteronarcys californica* and *Locusta migratoria,* Clubb et al. (1975) and Martoja et al. (1983) have caracterized a cadmium-binding glycoprotein which contains only one cysteinyl residue. This proteins seems to be typical of insects.

Cu-chelatins are another interesting class of metal-binding proteins which has been described in marine animals. Cu-chelatin, similar in molecular weight to metallohionein, has been first identified in mammalian tissues (Shapiro et al. 1961).

Cu-chelatin is very different from metallothionein in that it binds predominantly Cu and possesses a much lower cysteine content (8% as in the Cd-binding characterized from the american oyster by Ridlington and Fowler 1979). (Table 1). Ridlington et al. (1981) have described metal-binding proteins with amino acid composition similar to rat liver copper chelatin purified from sea lion kidney and liver, whale liver, crab tissues and fish liver.

In this regard, it is worth noticing the recent work of Viarengo et al. (1984) about the biochemical characterization of Cu-thioneins isolated from the tissues of *Mytilus galloprovincialis* exposed to the metal. Their results show that the use of classical techniques such as gel filtration and ion-exchange chromatography only partially purified the Cu-rich proteins suggesting the presence of Cu-chelatins. However, additional purification by denaturing polyacrylamide gel electrophoresis revelated that the completely purified Cu-rich protein showed characteristics common to most Cu-thioneins.

This suggests that some Cu-chelatins previously described in the literature could perhaps be in fact incompletely purified metallothioneins.

In any case, in marine animals, metallothioneins and metallothionein-like proteins reveal a wide distribution and can be considered as a storage mechanism of excess metals.

The detoxifying effect of metallothioneins in marine animals has been described by some authors (Bouquegneau et al. 1975; Bouquegneau 1979, Young and Roesijadi 1983; Roesijadi 1983) which have shown a correlation between the appearance of metallothioneins and an enhanced tolerance to potentially toxic metals such as mercury, cadmium, copper and zinc in fish, molluscs and polychaetes.

As suggested by Noël-Lambot et al. (1978), there may exist a threshold concentration for metals in tissues at which metallothionein is synthesized. Thus, comparing for example the Cd, Cu and Zn loads of *Patella vulgata* and *Littorina littorea* living in uncontaminated or contaminated natural waters or exposed to cadmium in the labortory, these authors have shown that if, in the laboratory, both species were able to synthesize high concentrations of metallothioneins, in the polluted field, only limpets contained metallothioneins in detectable amounts. Effectively, in the Bristol Channel, *Patella vulgata* contained much higher cadmium concentrations than *Littorina littorea* and an important load of Cd-binding thionein has been detected in animals with high levels of that metal. A more accurate study of the cadmium accumulation in limpets from the Bristol Channel was performed by the same authors (Noël-Lambot et al. 1980) and shows that there is a close correlation between the quantity of Cd bound to metallothioneins and the concentration of total Cd in the tissue from 14 ppm WW on while the cadmium in the high molecular weight proteins fractions of the cytosol remainted constant. Thus, considering the correlation between the age and the Cd content of the animals, it can be concluded that during long-term Cd intoxication of limpets under field conditions, Cd deposition and increase in metallothioneins in limpets were directly linked.

It is worth noticing now that the subcellular localization of metallothioneins is generally considered to be mainly in the soluble phase. A lysosomal localization of metallothioneins has also been reported by Porter (1974) in the new-born liver and in insects exposed to mercury by Jeantet et al. (1980) and Bouquegneau et al. (to be published). This lysosomal storage may represent the terminal phase of the cell cycle of metal detoxification; this leads us now to consider compartmentation storage processes of heavy metals by marine organisms.

2 Storage of Metals in Membrane-Limited Vesicles: Lysosomes and Spherocrystals

Binding to proteins is one way of metal storage common to vertebrate and invertebrate species. Another way is the immobilization of the metals in membrane-limited vesicles. Among these, lysosomes are also widely distributed in the animal kingdom but spherocrystals occur only in invertebrates.

Some authors have shown the presence of heavy metals in lysosomes of marine invertebrates: Walker (1977) has reported copper on lysosomes of the parenchyma cells of the barnacle *Balanus balanoides,* George et al. (1978) have described the immobilization of copper associated with sulphur and zinc with phosphorus in membrane-limited vesicles of the oyster *Ostrea edulis,* Ballan et al. (1982) have shown numerous metals such as Cd, Ag, Zn, Cd, and Fe concentrated in the lysosomes of digestive cells and amoebocytes of the digestive gland of *Pecten maximus* and Martoja et al. (1980) have described in the pore cells of connective tissure of the winkle *Littorina littorea* the presence of copper precipitated in the form of clusters of copper sulphide needles inside the lysosomes. Note that copper sulphide was identified by Raman spectrometry and that such a compound has been interpreted as being the result of haemocyanin degradation: proteins of the hemolymph are concentrated by coated vesicles of the pore cells and digested in lysosomes.

The presence of copper sulphide in residual bodies of pore cells which are not excreted from the animal is a factor which favours high copper loads in these animals. Indeed, recently, we have studied the total copper content and the copper sulphide content of another gastropod, *Murex brandaris,* in relation with the distribution and the aspect of sulphide copper cells which can be observed in the visceral mass of this animal (Bouquegneau and Martoja 1982).

Soft tissues of *Murex brandaris* contain high levels of copper compared with other species of gastropods, and there is a close correlation between age (or weight) and the copper content. Such a correlation appears only in gastropod species (neogastropods) in which copper sulphide can be detected. In *Murex brandaris,* for example, all the excess copper is located in the visceral mass of the animal where the most is under the copper sulphide form. The general aspect of the visceral mass is white when the animal is young and devoid of copper sulphide, whilst there are black marblings on the visceral mass of older animals which contain high levels of copper sulphide.

The naturally occurring copper sulphide is, in fact, contained in pore cells which are located between the glandular tubules and not enclosed in the digestive gland sensu stricto (Bouquegneau and Martoja 1982). A study of the visceral mass of *Murex brandaris* by secondary ion emission analysis has been performed and results are reported in Figs. 1 and 2.

Figure 1a shows an optical micrograph of the edge of the visceral hump in *Murex brandaris.* The tegument is located at the top left. Two digestive tubules are located at the bottom right. Between tegument and tubules, there is a connective tissue with black heaps corresponding to pore cells. Secondary ion emission micrographs of the next section are shown in Figs. 1b, c and d. Calcium (Fig. 1b) is mainly located at the level of spherocrystal cells of the digestive tubules. On the contrary, copper (Fig. 1c) and cadmium (Fig. 1d) are essentially located in pore cells.

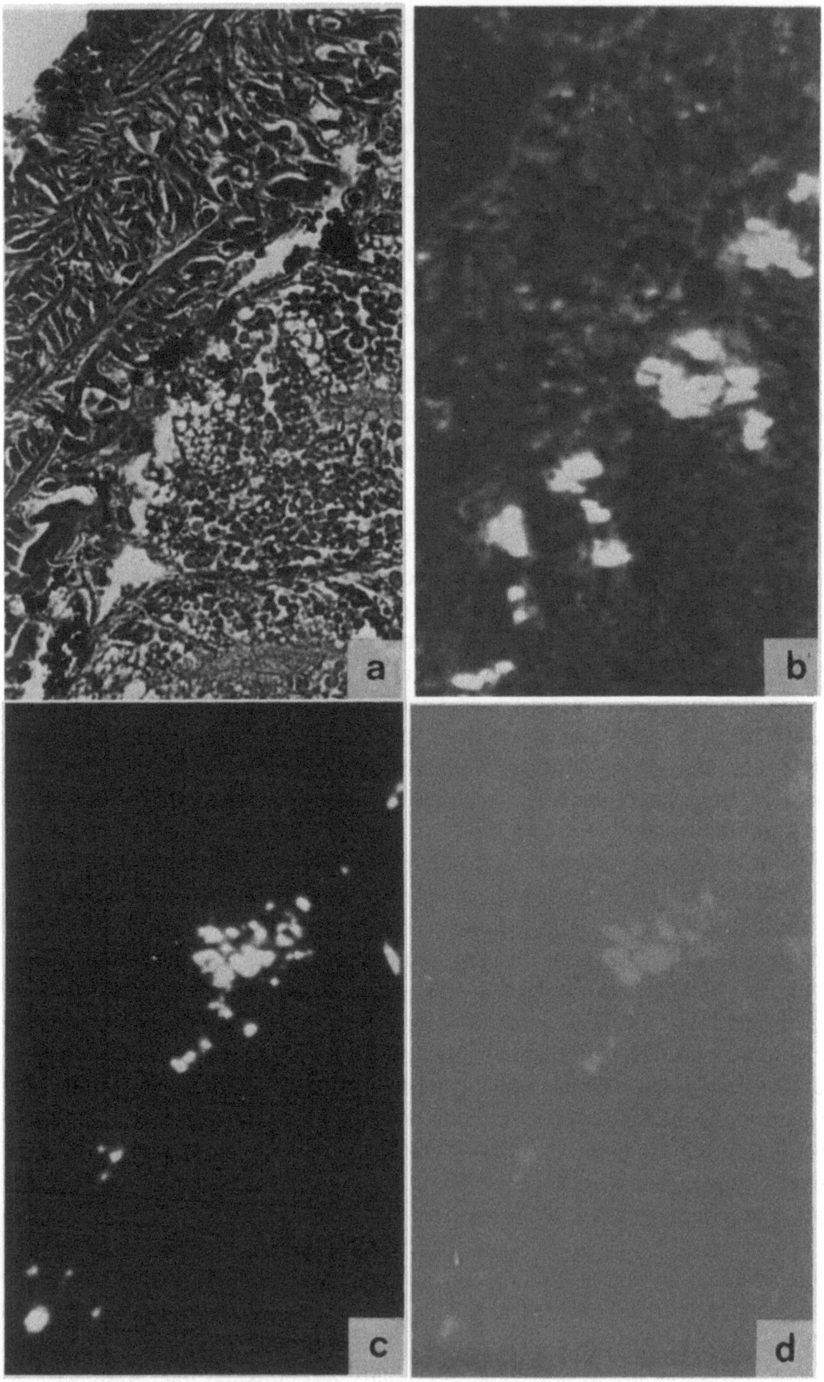

Fig. 1 a–d. (Legend see p. 154)

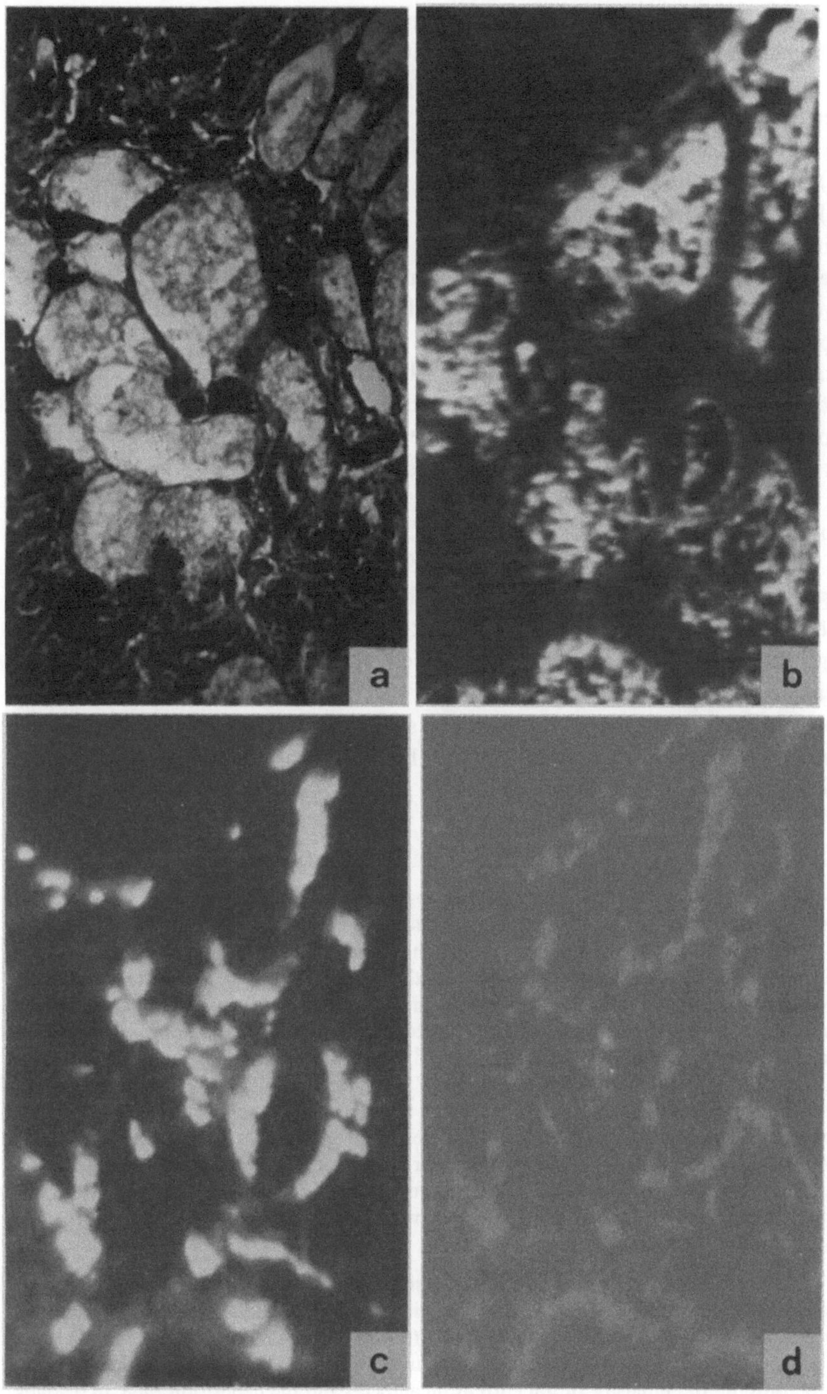

Fig. 2 a–d. (Legend see p. 154)

Figure 2a shows periarterial vesicular conjunctive cells of the digestive gland of *Murex brandaris.* These cells contain high calcium levels (Fig. 2b). Copper (Fig. 2c) is located between these vesicular cells in pore cells whilst cadmium (Fig. 2d) can be observed in both types of cells.

This view of cadmium and calcium in the periarterial vesicular conjonctive cells (Fig. 2) as well as the presence of calcium in high amounts in the digestive tubules (Fig. 1) leads us now to consider the second way of storing heavy metals in membrane-limited vesicles: spherocrystals.

Spherocrystals are characterized by concentric layers and a radial organization. Moreover, they differ from lysosomes by the fact that there is never in this structure any supply of hydrolytic enzymes. Whilst in lysosomes there is a degradation of organic matter, in spherocrystals there is a synthesis of an organic matter network on which the inorganic matter can be precipitated. In lysosomes, the degradation of sulphur containing amino acids may result in the formation of hydrogen sulphide (Fromageot 1951). This acid precipitates metals as sulphides which probably will not be able to be used again by the animal. On the contrary, in spherocrystals, precipitates are phosphates, carbonates or oxalates (see e.g. Martoja and Truchet 1983). They can rapidly incorporate Ca (Tompa and Watabe 1976) but can be dissolved again to supply oligoelements (for example calcium for the shell building or for vitellogenesis). Spherocrystals can also operate in the pH control (Sminia et al. 1977; Overnell 1983). For a general review of the role of calcium cells (containing spherocrystals) and of pore cells, see Richardot (1979).

Some authors have shown the presence in spherocrystals of metals other than calcium and magnesium. For example, Walker (1977) has shown by X-ray microprobe analysis the presence of zinc and phosphorus in granules made up of concentric layers in

---

Fig. 1 a–d. *Murex brandaris,* visceral hump. Histological sections of paraffinembedded material. a Optical micrograph (Erythrosin-Orangé G) showing tegument *(top left)* and two digestive tubules *(bottom right).* Between the tegument and the tubules stretches line of connective cells (pore cells) with black heaps of CuS. Just below, in the digestive tubules, lays spherocrystal cells which cannot be identified on this image (x 450; 1 cm = 22 μm). **b–d** Ion images (positive secondary emission) from another place with identical histological structure (same magnification). For each image, are given the mass of the isotope (ex. $^{40}Ca^+$), the intensity of the ionic current in amperes (ex. 10^{-14} A) and the exposition time of the photographic film in seconds. **b** $^{40}Ca^+$ (10^{-14} A, 20 s) Calcium is slightly emitted by whole tissues (in grey); pore cells cannod be identified, whereas calcium cells of the digestive tubules strongly emit Ca (white areas). **c** $^{63}Cu^+$ (10^{-14} A, 100 s). The very strong emission of copper is only located inside pore cells underlaying the tegument, just above the calcium cells of digestive tubules. **d** $^{114}Cd^+$ (5.10^{-17} A, 1200 s). As for copper, cadmium is emitted by pore cells, but with considerably smaller intensities, due to lower concentration and ionic yield

Fig. 2 a–d. *Murex brandaris,* periarterial region inside the visceral mass. **a** Optical micrograph, semi-thin section of resin-embedded material (Toluidine blue); this region contain black pore cells and grey vesicular conjonctive cells around a vessel (x 450; 1 cm = 22 μm). **b–d.** Ion images of another area, in the top of the visceral mass (same magnification). For each image are given the mass of the isotope, the intensity of ion current, in amperes and the deposition time, in seconds. **b** $^{40}Ca^+$ (6.10^{-14} A, 10 s). The numerous vesicles in the cytoplasm of vesicular cells emit intensively calcium. **c** $^{63}Cu^+$ (4.10^{-15} A, 100 s) Copper is strongly emitted by pore cells alone. **d** $^{114}Cd^+$ (7.10^{-17} A, 900 s) A light but characteristic emission of cadmium occurs from both cell types

the barnacle *Balanus balanoides*. Hignette (1979) has found manganese, aluminium, cobalt and zinc in spherocrystals of *Pinna mobilis* and *Tridacna maxima* kidneys. Carmichael et al. (1979) have shown the presence of trace metal content (Zn, Cd, Cu, and Cr) in *Argopecten irradians* and *A. gibbus* kidney concretions. Martoja et al. (1980) have shown in perivascular cells surrounding the aorta and arteries of *Littorina littorea* the presence of copper inside spherocrystals. Marshall and Talbot (1979) have shown that Cd was associated with S and frequently with P in membrane-bound vesicles of *Mytilus edulis* gills. Note that spherocrystals and lysosomes can also accumulate radionuclides, as discovered by Chassard-Bouchaud (1983) in the mussel *Mytilus edulis*.

It is worth noticing that, contrary to the fact we described above, that the storage of metals under a sulphide form in pore cells of the visceral mass is to be considered as a factor favouring high metal loads in the animal, granules (lysosomes or spherocrystals) which are built in tissues communicating with the outside of the animal may be eventually excreted as such into the urinary tract (George et al. 1980). In this regard however, the work by Martoja et al. (1982) on the contamination by cadmium of the sea water-adapted eel (Teleost) shows that nephropatic casts are produced in the kidneys and are mainly made of phosphate and oxalate of calcium, but, as they are devoided of cadmium, they do not contribute in this case to excreting the toxic metal.

We may conclude that there are essentially three ways by which marine animals can store heavy metals under a detoxified form: metallothioneins, lysosomes and spherocrystals. Recently, we have started the study of those three ways of storage in a single species, *Murex trunculus*, chosen for its high cadmium content even when caught in unpolluted waters.

3 Storage of Cadmium by *Murex trunculus*: Effect of Pollution and Seasonal Variations – Preliminary Results

Specimens of *Murex trunculus* were caught in October 1982 in Calvi's bay (Corsica) known for the absence of heavy metals contamination. Soft parts of this species contain high levels of cadmium (about 6 ppm WW) and the higher loads of cadmium are found in the visceral mass of the animals (Bouquegneau et al. 1983). When contaminated in captivity by cadmium added to the surrounding water, there is a significant increase of the cadmium content of the animals, significant but relatively low. Once again, the visceral mass is the most affected by the contamination. This organ has been examined both from the point of view of the biochemical and histological localization of cadmium. Table 2 shows the distribution of cadmium in the soluble and insoluble fractions of the digestive gland in October 82 compared to April 83. In October, Cd is equally distributed in each fraction. Few metallothioneins are detected in the soluble fractions of the visceral mass, so that all the cadmium of this fraction is bound to high molecular weight proteins. However, when contaminated in the laboratory, these animals synthetize metallothioneins which bind a large part of the accumulated cadmium (Bouquegneau et al. 1983).

In April, the total Cd content of the digestive gland does not change significantly but the relative distribution of the metal in the gland is quite different: cadmium

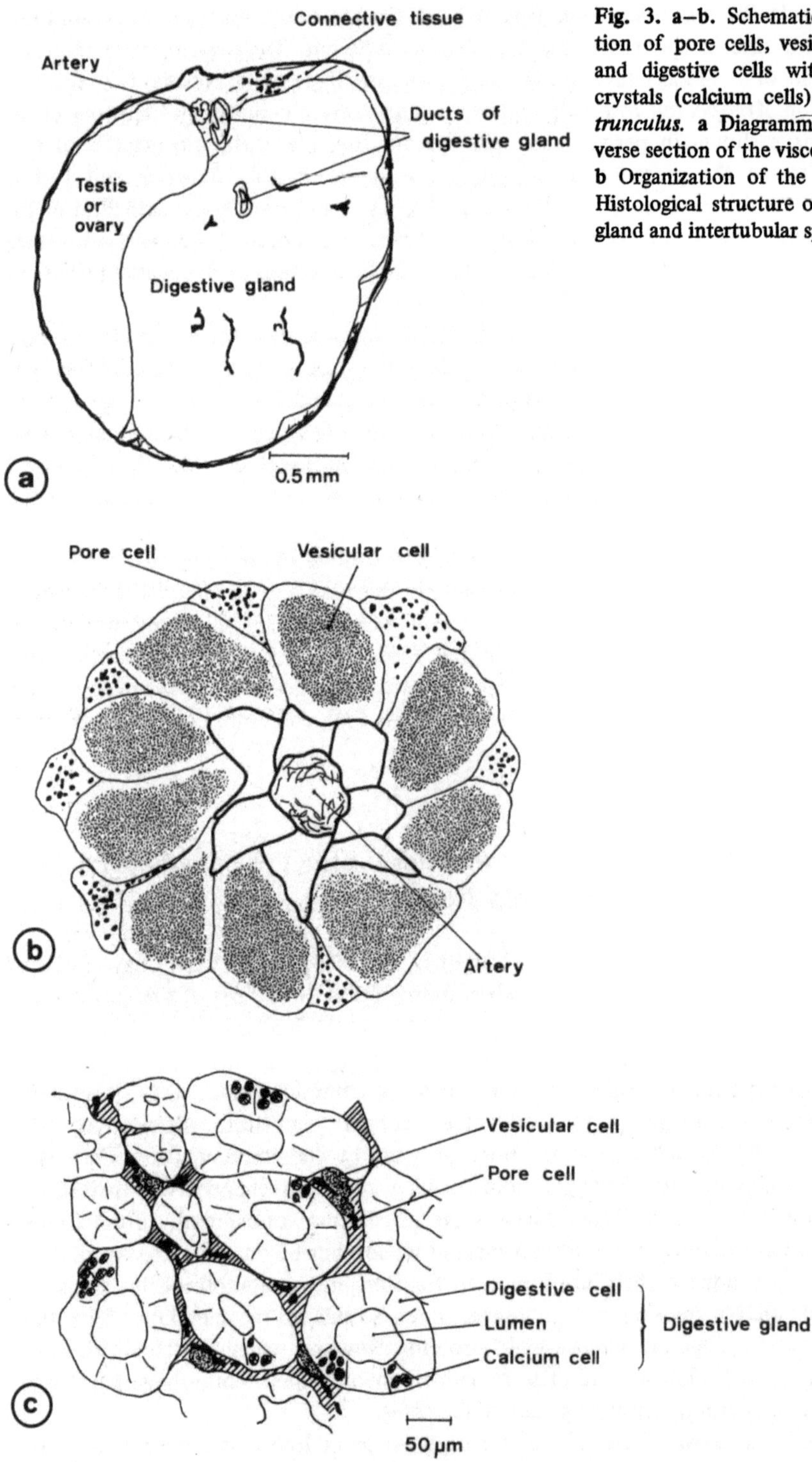

Fig. 3. a–b. Schematic localization of pore cells, vesicular cells and digestive cells with spherocrystals (calcium cells) in *Murex trunculus.* **a** Diagrammatic transverse section of the visceral hump. **b** Organization of the arteries. **c** Histological structure of digestive gland and intertubular spaces

Table 2. Biochemical localization of cadmium in the visveral mass of *Murex trunculus* caught in calvi's bay in October 1982 and in April 1983

Murex trunculus	Total cadmium (ppm WW)	Cadmium in soluble fraction		Cadmium in the insoluble fraction
Calvi's Bay	$\overline{m} \pm$ S.D.	HMWP-bound	MT-bound	
October 1982 n =7	20.2 ± 8.3	8.6 ± 5.1	0.7 ± 1.1	10.9 ± 4.0
April 1983 n = 5	25.9 ± 15.0	5.2 ± 2.8	16.0 ± 9.6	4.7 ± 3.1
Δ	+ 5.7 $P \cong 0.4$ N.S.	– 3.4 $0.2 > P > 0.1$	+ 15.3 $P < 0.01$	– 6.2 $P < 0.01$

HMPW = High Molecular Weight Proteins; MT = metallohionein

concentration decreases significantly in the insoluble fraction of the tissue and increases highly at the level of metallothioneins in the soluble fraction. These results show that there is a seasonal cycle of cadmium in the digestive gland of *Murex trunculus*.

Figure 3 shows a schematic view of the visceral hump of *Murex trunculus* which is composed of three distinct parts: the gonad, the digestive gland and the conncetive tissue of the arteries. The digestive gland is composed of calcium cells (which contain spherocystals) and of digestive cells. The intertubular spaces contain pore cells with copper sulphide in lysosomes and vesicular cells with spherocystals.

The connective tissue associated to the vascular system is quite particular and composed again of pore cells and vesicular cells. Secondary ion emission analysis has revealed (Martoja et al., in preparation) that either cadmium was associated with calcium of vesicular cells, or that it was mainly located in the pore cells, or was diffused in the whole tissue, depending on the individual. This suggests once again the existence of a cadmium cycling in the animal and that, perhaps, cadmium should be considered as a possible essential element for this species.

Whatever the case, these preliminary results show clearly that heavy metals can be stored in marine organisms in many ways and that these ways are depending not only on the species but also on environmental conditions such as the pollution level of the medium and seasonal variations.

References

Ballan-Dufrancais C, Jeantet AY, Halpern S (1982) Localisation intracellulaire par microanalyse X des métaux et de métalloides dans la glande digestive d'un mollusque bivalve *(Pecten maximus)*. Implication des processus de digestion. CR Acad Sci (Paris) 294 (Série III): 673–678

Bouquegneau JM (1979) Evidence for the protective effect of metallothionein against inorganic mercury injuries to fish. Bull Environ Contam Toxicol 23: 218–219

Bouquegneau JM, Gerday Ch, Distèche A (1975) Fish mercury-binding thionein related to adaptation mechanisms. FEBS Lett 55: 173–177

Bouquegneau JM, Martoja M (1982) La teneur en cuivre et son degré de complexation chez quatre Gastéropodes marins. Données sur le cadmium et le zinc. Oceanol Acta 5: 219–228

Bouquegneau JM, Martoja M, Truchet M (1983) Localisation biochimique du cadmium chez *Murex trunculus* (L.) (Prosobranche Néogastéropode) en milieu naturel non pollué et après intoxication expérimentale. CR Acad Sci (Paris) 296 (Série III): 1121–1124

Carmichaël NG, Squibb KS, Fowler BA (1979) Metals in the molluscan kidney: a comparison of two closely related bivalve species (Argopecten), using x-ray microanalysis and atomic absorption spectroscopy. J Fish Res Board Can 36: 1149–1155

Casterline JL JR, Yip G (1975) The distribution and binding of cadmium on oyster, soybean, and rat liver and kidney. Bull Environ Contam Toxicol 3: 319–329

Chassard-Bouchaud C (1983) Rôle des loysosomes et des sphérocristaux dans le phénomène de concentration de l'uranium chez la moule *Mytilus edulis* (L.). Microanalyse par spectrographie des rayons X. CR Acad Sci (Paris) 296 (Série III): 581–586

Clubb RW, Lords JL, Gaufin AR (1975) Isolation and characterization of a glycoprotein from the stonefly *Pteronarcys californica,* which binds cadmium. J Insec Physiol 21: 53–60

Flos R, Hidalgo JM (1983) Cd metallothionein in elasmobranchs: its presence in the dogfish. Abstracts of the 5th Conference of the European Society for Comparative Physiology and Biochemistry, September 5–8, Taormina (Italy), pp 94–95

Frankenne F, Noël-Lambot F, Disteche A (1980) Isolation and characterization of metallothioneins from cadmium-locaded mussel *Mytilus edulis.* Comp Biochem Physiol 66C: 179–182

Fromageot C (1951) The enzymes, chemistry and mechanism of action. Sumner JB and Myrbäch K (eds), vol 2, part 1. Academic, p 620

George SG, Carpene E, Coombs TL, Overnell J, Youngson A (1979) Characterization of cadmium-binding protein from mussels *Mytilus edulis* (L.) exposed to cadmium. Biochim Biophys Acta 580: 225–233

George SG, Pirie BJS, Chenye AR, Coombs TL, Grant PT (1978) Detoxication of metals by marine bivalves: an ultrastructural study of the compartmentation of copper and zinc in the oyster *Ostrea edulis.* Mar Biol (NY) 45: 147–156

George SG, Pirie BJS, Coombs TL (1980) Isolation and elemental analysis of metal-rich granules from the kidney of the scallop *Pecten maximus* (L.). J Exp Mar Biol Ecol 42: 143–156

Hignette M (1979) Composition des concrétions minérales contenues dans les reins de deux mollusques lamellibranches: *Pinna mobilis* (L.) et *Tridacna maxima* (Röding). CR Acad Sci (Paris) 289 (Série D): 1069–1072

Howard AG, Nickless G (1977) Heavy metal complexation in polluted molluscs. I. Limpets *(Patella vulgata* and *Patella intermedia).* Chem Biol Interactions 16: 107–114

Irons RD, Smith JC (1977) Isolation of a non-thionein copper-binding protein from liver of copper-injected rats. Chem Biol Interactions 18: 83–89

Jeantet AY, Ballan-Dufrancais C, Ruste J (1980) Quantitative electron probe microanalysis on insects exposed to mercury. II. Involment of the lysosomal system in detoxification processes. Biol Cell 39: 325–334

Jennings JR, Rainbow PS, Scott AG (1979) Studies on the uptake of cadmium by the crab *Carcinus maenas* in the laboratory. II. Preliminary investigation of cadmium-binding proteins. Mar Biol (NY) 50: 141–149

Kojima Y, Kagi JHR (1978) Metallothionein. Trends Biochem Sci 3: 90–93

Lyon R, Taylor M, Simkiss K (1983) Metal-binding proteins in the hepatopancreas of the crayfish *(Austropotamobius pallipes).* Comp Biochem Physiol 74C: 51–54

Margoshes M, Vallee BL (1957) A cadmium protein from equine kidney cortex. J Am Chem Soc 79: 4813

Marshall AT, Talbot V (1979) Accumulation of cadmium and lead in the gills of *Mytilus edulis:* X-ray microanalysis and chemical analysis. Chem Biol Interactions 27: 111–123

Martoja M, Truchet M (1983) Données analytiques sur les concrétions du tissu conjonctif de quelques gastéropodes d'eau douce. Malacologia 23: 333–349

Martoja M, Vu Tan Tue, Elkaim B (1980) Bioaccumulation du cuivre chez *Littorina littorea* (L.) (Gastéropode prosobranche): signification physiologique et écologique. J Exp Mar Biol Ecol 43: 251–270

Martoja R, Bouquegneau JM, Verthé C (1983) Toxicological effects and storage of cadmium and mercury in an insect *Locusta migratoria* (Orthoptera). J of Invertebr Pathol 42: 17–32

Martoja R, Truchet M, Bouquegneau JM (1982) Une néphrophatie provoquée par le cadmium chez l'anguille adaptée a l'eau de mer (Téléostéen). CR Acad Sci (Paris) 295 (Série III): 369–374

Noël-Lambot F (1976) Distribution of cadmium, zinc and copper in the mussel Mytilus edulis. Existence of cadmium-binding proteins similar to metallothioneins. Experientia (Basel) 32: 324–326

Noël-Lambot F (1980) Cadmium accumulation correlated with increase in metallothioneins concentration in the limped *Patella caerulea*. In: Gilles R (ed) Animals and environmental fitness. Proceedings of the Ist Meeting of the European Society of Comparative Physiology and Biochemistry. Pergamon, New York

Noël-Lambot F, Bouquegneau JM, Frankenne F and Distèche A (1978) Le rôle des métallothionéines dans le stockage des métaux lourds chez les animaux marins. Rev Int Océanogr Méd XLIX: 13–20

Noël-Lambot F, Bouquegneau JM, Frankenne F, Disèche A (1980) Cadmium, zinc and copper accumulation in limpets *(Patella vulgata)* from the Bristol Channel with special reference to metallothioneins. Mar Ecol Prog Ser 2: 81–89

Olafson RW, Thompson JAJ (1974) Isolation of heavy metal binding proteins from marine vertebrates. Mar Biol (NY) 28: 83–86

Olafson RW, Sim RG, Boto KG (1979a) Isolation and chemical characterization of the heavy metal-binding protein metallothionein from marine invertebrates. Comp Biochem Physiol 62B: 407–416

Overnell J (1983) Protein and oxalate in mineral granules from the kidney of *Pecten maximus* (L.). J Exp Mar Biol Ecol 52: 173–183

Overnell J, Coombs TL (1979) The isolation of plaice metallothionein. Biochem J 183: 277–283

Overnell J, Trewhella E (1979) Evidence for the natural occurrence of (cadmium-copper) metallothionein in the crab *Cancer pagurus*. Comp Biochem Physiol 64C: 69–76

Porter H (1974) The particulate half-cystine-rich copper protein of newborn liver. Relationship to metallothionein and subcellular localization in non-mitochondrial particles possibly representing heavy lysosomes. Biochem Biophys Res Comm 56: 661–668

Richardot M (1979) Calcium cells and groove cells in calcium metabolism in the freshwater limpet *Ferrissia wautieri* (Basommatophora Ancylidae). Malacol Rev 12: 67–78

Ridlington JW, Chapman DC, Goeger DE, Whanger PD (1981) Metallothionein and Cu-chelatin: characterization of metal-binding proteins from tissues of four marine animals. Comp Biochem Physiol 70B: 93–104

Ridlington JW, Fowler BA (1979) Isolation and partial characterization of a cadmium-binding protein from the American oyster *(Crassostrea virginica)*. Chem Biol Interactions 25: 127–138

Roesijadi G (1981) The significance of low molecular weight, metallothionein-like proteins in marine invertebrates: current status. Mar Environ Res 4: 167–179

Roesijadi G (1983) Enhanced mercury tolerance in *Mytilus edulis*: influence of Cu, Zn and Cd pre-exposure and relationship to metal-binding proteins. Abstracts of the 5th Conference of the European Society for Comparative Physiology and Biochemistry, September 5–8, Taormina (Italy), pp 102–103

Roesijadi G, Hall RE (1981) Characterization of mercury-binding proteins from the gills of marine mussels exposed to mercury. Comp Biochem Physiol 70C: 59–64

Sadler PJ, Bakka A, Beynon PJ (1978) ^{113}Cd nuclear magnetic resonance of metallothionein. FEBS Lett 315: 318

Shapiro J, Morell AG, Scheinberg IH (1961) A copper proteins of human liver. J Clin Invest 40: 1081

Simkiss K, Mason AZ (1983) Metal Ions: Metabolic and Toxic Effects. In: The Mollusca, vol. 2. Environmental biochemistry and physiology. Academic, New York

Sminia T, De With ND, Bos JL, Van Nieuwmegen ME, Witter MP, Wondergem J (1977) Structure and function of the calcium cells of the freshwater Pulmonate snail *Lymnaea stagnalis*. Neth J Zool 27: 195–208

Tompa A, Watabe N (1976) Calcified arteries in a gastropod. Calc Tissue Res 22: 159–179

Viarengo A, Pertica A, Mancinelli G, Palermo S, Orunesu M (1980) Rapid induction of copper-binding proteins in the gills of metal-exposed mussels. Comp Biochem Physiol 67C: 215–218

Viarengo A, Pertica M, Mancinelli G, Zanicchi G, Bouquegneau JM, Orunesu M (1984) Biochemical characterization of Cu-thioneins isolated from the tissues of mussels exposed to the metal. Mol Physiol 5: 41–52

Walker G (1977) "Copper" granules in the barnacle *Balanus balanoides*. Mar Biol (NY) 39: 343–349

Winge DR, Geller BL, Garvey J (1981) Isolation of copper-thionein from rat liver. Arch Biochem Biophys 208: 160–166

Young JS, Roesijadi G (1983) Reparatory adaptation to copper-induced injury and occurrence of a copper-binding protein in the polychaete *Eudistylia vancouveri*. Mar Pollut Bull 14: 30–32

Biochemical Differences in Field Populations of the Common Mussel *Mytilus edulis* L. Exposed to Hydrocarbons: Some Considerations of Biochemical Monitoring

D.R. LIVINGSTONE[1]

1 Introduction

The main aims of studies of the biochemical responses of marine invertebrates to pollution are the detection of adverse effects and the elucidation of mechanisms by which the organisms might function in and possibly adapt to the new environment. Techniques which measure the biological effects of pollution are considered critical to any programme of environmental impact assessment (Cairns and Van der Schale 1980; Rosenberg et al. 1981) and in recent years a number of so-called indices of stress, measuring effects at various levels of organization (biochemical and cytochemical, physiological, ctyological etc.), have been identified, their practicability considered (e.g., Bayne et al. 1980; Lee et al. 1980; Livingstone 1982a) and, in some cases, put into routine use, e.g., physiological scope for growth (Bayne et al. 1979, 1981, 1982) and cytochemical lysosomal latency (Moore 1980, 1982; Bayne et al. 1981). Biochemical indices offer the potential of sensitivity and ease of measurement but in particular that of specificity in that certain biochemical systems may respond only to one type of environmental stressor and therefore aid in its identification in a complex environmental situtation; examples of such "specific indices of stress" (as opposed to "general indices of stress") are metallothioneins responding to heavy metals and the mixed function oxygenase (MFO) system responding to organic xenobiotics.

Problems still exist in the application of biochemical studies to biological monitoring and the aim of this paper is to discuss some of these in relation to a field study of populations of the common mussel, *Mytilus edulis,* from a clean environment and from the vicinity of a large oil terminal, the Sullom Voe Oil Terminal (SVOT), in the Shetland Islands; some laboratory data are also presented. The three enzyme systems studied are the glycolytic regulatory enzyme phosphofructokinase (PFK) (E.C.2.7.1.11) of the posterior adductor muscle, glucose-6-phosphate dehydrogenase (G6PDH) (E.C.1.1.1.49) in blood cells and the microsomal MFO system in the digestive gland (hepatopancreas). Whereas PFK responds to a number of stressors (Widdows et al. 1982; Blackstock and Filion-Myklebust 1983), G6PDH and the MFO system are indicated to be more specific for hydrocarbons and other organic xenobiotics (Moore et al. 1980; Livingstone and Farrar 1984).

1 Natural Environment Research Council, Institute for Marine Environmental Research, Prospect Place, The Hoe, Plymouth, PLI 3 DH, U.K.

Toxins, Drugs, and Pollutants in Marine Animals
(ed. by Bolis et al.)

2 Some Considerations of Biochemical Monitoring

The problems of biochemical monitoring essentially fall into two categories, the practical considerations of the detection of biochemical changes or differences against the background variability, and the interpretation or significance of such observed differences. The practical considerations are similar to those that have been discussed for some physiological (rates of oxygen consumption, the scope for growth) and cellular (the cytochemical latency of a lysosomal enzyme) processes (Bayne et al. 1981), but in particular, for biochemical processes include seasonal variability.

The ecology of *Mytilus* and many other marine invertebrates involves an annual cycle of the storage of food reserves and its subsequent utilization for maintenance energy and gametogenesis. As a consequence of this and other factors such as changing environmental temperature, most aspects of metabolism show a regular seasonal pattern of change. These occur not only in storage and reproductive tissues (e.g., the mantle) but also in other tissue types (e.g., the posterior adductor muscle) of *Mytilus*. Seasonal changes in *Mytilus* have been observed in concentrations of biochemical substrates, e.g., glycogen, lipid, protein and free amino acids (Zandee et al. 1980), nucleic acids (Thompson 1972), total free sugars (Bayne 1973), free glucose (Zaba 1981; Livingstone and Clarke 1983), glycolytic intermediates (Ebberink and De Zwaan 1980) and phosphoarginine (Zurburg and Ebberink 1981), and in rates of processes, e.g., nitrogen excretion (Bayne and Scullard 1977), osmoregulation (Livingstone et al. 1979), anaerobic metabolism (Ahmad and Chaplin 1979; Zandee et al. 1980; De Zwaan et al. 1983a), pentose phosphate pathway activity and glucose utilization (Zaba et al., 1981). Not surprisingly, because such changes will be effected by flux alterations through enzyme-catalyzed reactions, seasonal changes in enzyme activities and kinetic characteristics are also evident, e.g., pyruvate kinase (E.C.2.7.1.40) (Livingstone 1975), glucose-6-phosphate dehydrogenase (Livingstone 1981), glycogen synthetase (E.C.2.4.1.11) (Gabbott et al. 1979) and hexokinase (E.C.2.7.1.1.) (Livingstone and Clarke 1983). The importance of such variability is not only that it makes the detection of biochemical differences more difficult, but that differences could exist between two populations of mussels simply because they are at different stages of the reproductive cycle and not as a result of any direct effects of pollution. Such possibilities can be greatly reduced by sampling populations at the same time, and eliminated on the basis of a knowledge of the seasonal cycle of the particular parameter and the current reproductive state of the animal, e.g., for scope for growth see Bayne et al. (1981).

The interpretation of the in vivo significance of an observed change in, say, in vitro enzyme activities can be approached either empirically or by a fundamental understanding of the system or by some combination of the two, e.g., for lysosomal latency see Moore (1980); Bayne et al. (1981); Widdows et al. (1981, 1982). The empirical approach involves establishing the normal physiological range of values for the parameter based on baseline field surveys and experimental laboratory exposures to pollutants and other stressors, in which the animals are well characterized in terms of other measures of condition or health. Deviations of the biochemical parameter outside the "range of normality" would then be considered significant (Weber 1963). The fundamental approach should be more reliable but is likely to be more demanding and difficult to fully realise since, for example, controversy and debate still exist as to the effect that altered enzyme activities have on flux through metabolic pathways (Kacser and Burns 1979; Newsholme and Crabtree 1981). Ultimately, at least with respect to pollution studies, the answer probably lies in the demonstration of an altered in vivo flux or metabolic consequence, e.g., as for ATPase and osmoregulation (Kinter et al. 1972; Caldwell 1974; Schmidt-Nielsen et al. 1977). Given such problems, an approach that may have some general application in pollution studies is the measurement of the specific activities of enzymes that represent the maximum potential fluxes through particular metabolic pathways (Newsholme and Start 1973), e.g., phosphorylase (E.C.3.4.1.1) as a measure of the maximum potential rate of glycogenolysis, hexokinase for glucose utilization and phosphofructokinase for glycolysis. The correlations between the maximal specific activities of these enzymes and the maximum in vivo fluxes have been demonstrated experimentally (Newsholme and Crabtree 1973; Crabtree and Newsholme 1975) and the approach has the advantage that the enzyme changes can then at least be interpreted in terms of an altered potential of the pathway.

3 Field and Laboratory Studies

The three enzymes and their tissue locations were chosen for study for various reasons. PFK is a key enzyme in the regulation of flux through the glycolytic pathway which supplies carbon-skeleton for both aerobic and anaerobic energy-producing reactions: the posterior adductor muscle functions to open and close the mussel's valves so isolating the animal from the environment. G6PDH is an inducible enzyme catalyzing a reaction that produces NADPH which is a necessary substrate for the MFO reactions and is also involved in other detoxication processes through glutathione metabolism: blood cells were originally chosen because they are particularly active in the likely MFO-associated enzyme activity NADPH–neotetrazolium reductase (see Moore et al. 1980). The MFO system is an apparently universally distributed inducible enzyme system that catalyses the mono-oxygenation (hydroxylation) of organic xenobiotics rendering them more water soluble for excretion: the digestive gland in *Mytilus* is the primary site of hydrocarbon uptake (Widdows et al. 1983) and the primary tissue location of the MFO system (Livingstone and Farrar 1984).

The data presented are largely field data from the SVOT study with some supportive laboratory data for G6PDH and the MFO system. Following the announcement in 1972 of the discovery of oil below the North Sea, 100 miles north-east of Shetland, construction began in 1974 of the SVOT on the Shetland mainland and was largely completed in 1979 when the piping of oil ashore from the oil fields began. No complex refining of crude oil is carried out at SVOT and the main effluent discharged into the aquatic environment is ballast water that has been removed from the incoming tankers and first treated (for further details of the chemical and physical aspects of SVOT and its environment see the collected papers of the Proc. Roy. Soc. Edinb., 80B, 1–367 1981). Mussel collections were started in 1977 and subsequently carried out at approximately annual intervals (see Figures for further details

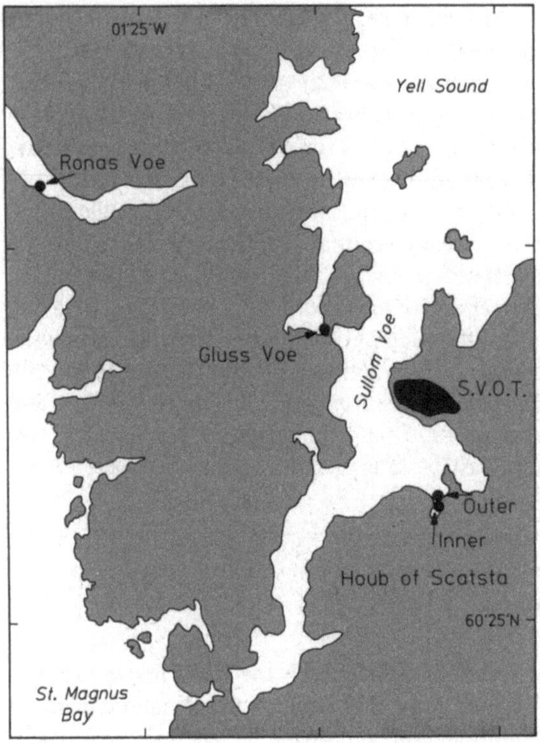

Fig. 1. Location of *M. edulis* populations in the Shetland Islands field study: Ronas Voe (G.R.292810); Gluss Voe (G.R.368775); Inner Houb of Scatsta (G.R.396732); Outer Houb of Scatsta (G.R. 396733). S.V.O.T. represents the site of the Sullom Voe oil terminal

of sampling times). Four native populations were sampled and selected on the basis of possible environmental contamination from the oil terminal (Fig. 1). Ronas Voe is isolated from Sullom Voe, Gluss Voe is in the same body of water but some distance from SVOT and the Houb of Scatsa populations are in the immediate vincinity of SVOT. The two Houb populations are separated by a sandbank with the Inner population being located closer inshore.

3.1 Materials and Methods

Mussels (4.5 to 5.5 cm length) were collected from the various sites while still submerged and quickly returned dry to a laboratory within SVOT for immediate processing. Blood samples were obtained by hypodermic syringe from the posterior adductor muscle sinus of individual mussels and the blood cells isolated and assayed for G6PDH as described in Moore et al. (1980). Different mussels were taken and the poterior adductor muscles dissected out, processed and assayed for PFK as described in Widdows et al. (1982). Digestive glands were also removed, damp-dried, frozen in liquid nitrogen and stored at $-70^{\circ}C$ before subsequent analysis for activities of the MFO system and its components at I.M.E.R., Plymouth. The microsomal fraction was prepared and assayed for G6PDH, NADH-ferricyanide reductase (NADH-FERRIRED and NADH- and NADPH-cytochrome c reductase [NAD(P)H- CYTCRED] activities as described in Livingstone and Farrar (1984) Benzo[a]pyrene hydroxylase (BPH) was assayed radiometrically by the method described in Binder and Stegeman (1980). All enzyme assays were carried out at 25°C. Protein was measured by the method of Lowry et al. (1951). Details of the numbers of samples taken are given in the text.

In addition to the biochemical analyses described above, both general measurements of animal condition (scope for growth, lysosomal latency and reproductive state) and specific aspects (tissue hydrocarbon analysis and NADPH- neotetrazolium reductase activity) were also carried out on the field mussels by other workers from I.M.E.R. (see Widdows et al., 1981 and the text for details) and some of these data are referred to.

Statistical analysis was by one-way or two-way analysis of variance between groups of values and, unless stated otherwise, a level of significance of $\leqslant 0.05$ was accepted. Details of laboratory studies and experiments are given in the text.

3.2 Results and Discussion

The early results of the field study, from October 1977 to October 1978, have been published in Widdows et al. (1981). Stereological assessment of the four populations showed typical reproductive cycles, similar to those normally observed for mussels from temperate waters, with no major differences between populations. Analysis of tissue hydrocarbons showed very low levels with no differences between populations up to October 1978 but with a marked accumulation, up to an order of magnitude higher, in the two Houb populations by July 1982 (Table 1). Lysosomal latency and scope for growth indices indicated that all four populations were healthy but with the two Houb populations being slightly stressed relative to Gluss Voe and Ronas Voe: latency of lysosomal hexosaminidase (E.C.3.2.1.30) was slightly lower in the Houb population on all sampling occasions (Moore pers. commun.), whereas the small relative decline in scope for growth in the Houb populations occurred from

Table 1. Tissue hydrocarbon concentrations of some Shetland populations of *M. edulis* in July 1982

Population	2-aromatic rings (2,3-dimethyl naphthalene equivalents)	3-aromatic rings (l-methylphenanthrene equivalents)
Ronas Voe	0.04 ± 0.01	0.03 ± 0.01
Gluss Voe	0.08 ± 0.01	0.03 ± 0.01
Inner Houb	0.40 ± 0.02	0.17 ± 0.02
Outer Houb	0.67 ± 0.16	0.25 ± 0.06

Data by kind permission of P. Donkin. Concentrations are in $\mu g\ g^{-1}$ wet weight and are mean and range. The total tissues of five mussels were pooled per analysis. The hydrocarbons were extracted by steam distillation and analysed by high performance liquid chromatography as described in Donkin and Evans (1984).

October 1978 onwards (Widdows pers. commun.). Against this assessment of the mussel populations' general condition, the biochemical data can be considered.

3.2.1 Posterior Adductor Muscle Phosphofructokinase

PFK measurements were started only in 1980 but since then the specific activities were consistently higher in the Houb populations than in Ronas Voe and Gluss Voe (Fig. 2). Given the absence of data before 1980, it might be argued that the differences in enzyme activity could be genotypic rather than phenotypic but the close location of the Gluss Voe population to the Houb of Scatsa and its distance from Ronas Voe would seem to argue against this possibility.

The increased potential glycolytic flux indicated in the posterior adductor muscles of the Houb animals can paradoxically be interpreted in terms of an increased requirement for either aerobic or anaerobic derived energy production. Whole animal respiration rates of mussels from both Houb populations were higher than those of Gluss Voe or Ronas Voe animals from October 1978 onwards (Widdows pers.commun.) and this would necessitate an increased flow of carbon to the Krebs cycle, at least in some tissues. The elevation of whole animals respiration rate with the laboratory exposure of mussels to the water-accommodated fraction (WAF) of North Sea crude oil has been observed and this was accompanied by an increase in PFK activity of the gills, although not of the posterior adductor muscle (Widdows et al. 1982). A more likely explanation for the higher PFK activities of the Houb animals, however, is an increased requirement for anaerobic metabolism. Two main types of anaerobic pathway, the succinate and strombine pathways, are used in the posterior adductor muscle, and both require glycolytic pyruvate (Livingstone 1982b). The succinate path-

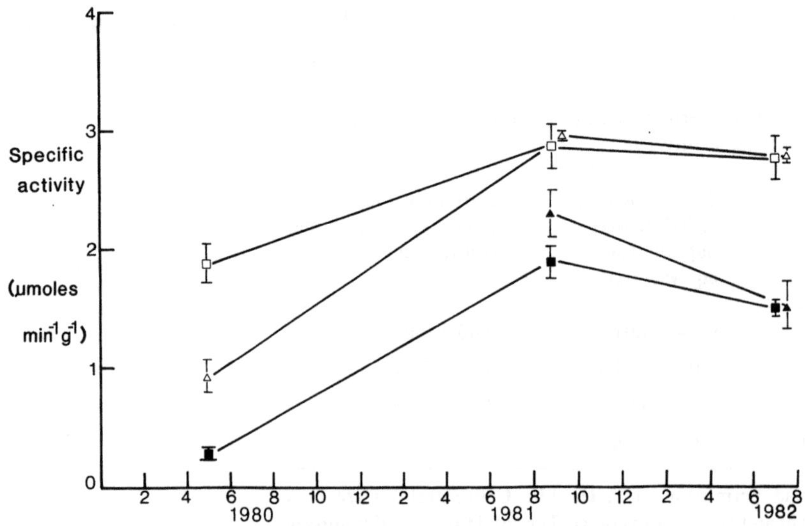

Fig. 2. Posterior adductor muscle phosphofructokinase activities of populations of *M. edulis.* Values are means ± SEM (n = 5 or 6; each sample is the pooled tissues of three mussels). ■: Gluss Voe; ▲: Ronas Voe; △: Outer Houb; □: Inner Houb

way gives rise to low rates of energy production and operates in the long term during valve closure. In contrast, the strombine pathway gives rise to relatively highe rates of energy production and operates during both the early stages of valve closure and the subsequent period of physiological activity following valve re-opening (De Zwaan et al. 1983b) and possibly is associated as much with valve movements as with survival during valve closure (De Zwaan pers. commun.). The higher PFK activities in the adductor muscles of Houb animals may therefore be the result of an increased requirement for valve closure and valve movements in response to a hostile environment in which there is some hydrocarbon effluent but also a high total particulate load (Widdows et al. 1981), the latter presumably resulting from the local industrial and construction activity around the Houb of Scatsa.

Changes in PFK activity were also observed over time with the summer and autumn values of 1981 and 1982 being high and the spring values of 1980 being low (Fig. 2). This difference is explainable in terms of the marked seasonal pattern of change that occurs in the specific activity of posterior adductor muscle PFK. The activity is high in autumn, declines in winter and remains low in spring before increasing again in summer (Churchill and Livingstone unpublished data).

3.2.2 Blood cell Glucose-6-Phosphate Dehydrogenase

The specific activity of G6PDH of blood cells has been observed to increase when mussels were either injected with polyaromatic hydrocarbons or exposed to such compounds in sea water (Moore et al. 1980). In the latter case the response also included an increase in the number of blood cells circulating in the haemolymph; this was reflected in a marked increase in blood G6PDH "total activity", i.e., the G6PDH activity per ml blood. MFO-associated enzyme activities such as benzo[a]pyrene hydroxylase (Livingstone and Farrar (1984), aldrin epoxidase and the apparently inducible NADPH- neotetrazolium reductase (Moore et al. 1980) are also present in blood cells and, given the theoretical bases discussed earlier, indicate that the changes in G6PDH activity were possibly a specific response to hydrocarbons.

In order to examine further this hypothesis an experiment was carried out in which mussels were exposed to hydrocarbons or to a general stressor (temperature) or to a combination of the two (Table 2). Marked changes were seen in blood G6PDH total activity which increased in response to both hydrocarbons and temperature and increasingly so far higher WAF levels. The indicated increase in the number of haemolymph blood cells was very obvious at high WAF and the WAF/ temperature combination with the blood becoming visibly whiter. In contrast, a sinificant increase in blood cell-specific activity was only observed in response to the combination of WAF and temperature. To an extent the results of previous experiments were maintained with the observed trend of an increase in the mean G6PDH specific activity with low and high WAF but not with temperature, although the high individual variability prevented the changes being statistically significant. Considering all the data of this and previous experiments, the indications therefore are that changes in blood G6PDH total activity represents a general response to different stressors effected through an increase in circulating blood cells, whereas the change in blood cell specific activity is more likely to be a specific response to hydrocarbons.

Table 2. Responses of glucose-6-phosphate dehydrogenase activity in the blood cells of *M. edulis* to the water-accommodated fraction (WAF) of North Sea crude oil and to temperature

Condition	Total activity (nmol min^{-1} ml^{-1} blood)	Specific activity (nmol min^{-1} mg^{-1} blood cell protein
Control	11.5 ± 2.7	47.9 ± 5.3
Low WAF (50 $\mu g\ l^{-1}$)	21.9 ± 3.5[a]	50.8 ± 6.5
High WAF (600 $\mu g\ l^{-1}$)	67.9 ± 9.6[a]	57.1 ± 4.8
Increased temperature (+ 10°C)	27.7 ± 4.8[a]	45.9 ± 3.5
Low WAF and increased temperature (50 $\mu g\ l^{-1}$ / + 10°C)	72.3 ± 15.6[a]	74.5 ± 4.9[a]

Mussels from the estuary of the River Erme, Plymouth were fed a maintenance ration and held for 2 weeks at each condition in flowing sea water. Levels of WAF and hydrocarbon analysis of sea water were achieved by the methods described in Widdows et al. (1982). Individual mussels were sampled and values are means ± SEM (n = 6).
[a] Significantly different from Control condition

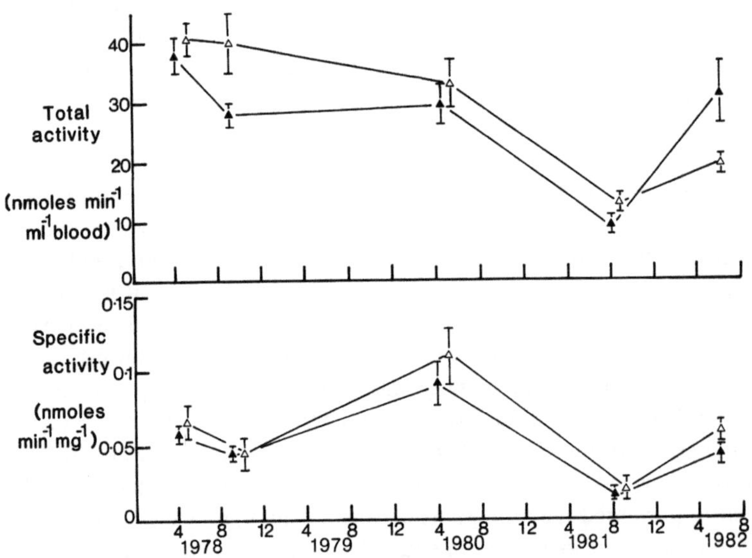

Fig. 3. Blood cell glucose-6-phosphate dehydrogenase activities of populations of *M. edulis.* Values are means ± SEM (n = 6,8 or 10; each sample is from an individual mussel). ▲: Gluss Voe; △: Outer Houb

The field study results are given in Fig. 3 for Gluss Voe and Outer Houb which were the only populations continously sampled. No significant differences between populations were seen at any one time but over time both blood G6PDH total activity (1978 to 1981) and blood cell-specific activity (1978 to 1982) were slightly higher in the Outer Houb mussels ($P<0.1$). The differences were most obvious in total activity and given that they were greatest in late 1978, before the arrival of tankers began, would suggest that they represent, like the changes in posterior adductor muscle PFK, a response to a generally hostile environment. In contrast, no time-course was obvious for the specific activity differences and although they may be suggestive of a specific response to hydrocarbons, the results are inconclusive. The changes in G6PDH total activity after 1981 represent a change in trend but require future data over time for interpretation.

The patterns of change over time were different for specific activity and total activity and, although no other seasonal information is available for blood cell G6PDH, the nature of the differences is as might be expected. A regular seaonal pattern of change was indicated for the specific activity of blood cells (values were consistently lower in autumn and higher in spring) but not for the number of blood cells (total activity) which can change rapidly in the short term due to their movement between the haemolymph and the tissues.

3.2.3 Digestive Gland MFO-Associated Enzyme Activities

Up until a few years ago, the presence of a cytochrome P-450 mediated MFO system in bivalve molluscs was regarded as doubtful. Recently, however, a number of studies have demonstrated its existence in these organisms and shown it to be primarily localized in the microsomal fraction of the digestive gland (see Livingstone and Farra 1984). Detectable activities include MFO activities such as BPH, (Fig. 4 and Table 3) and components or associated enzymes of the MFO system such as, re-

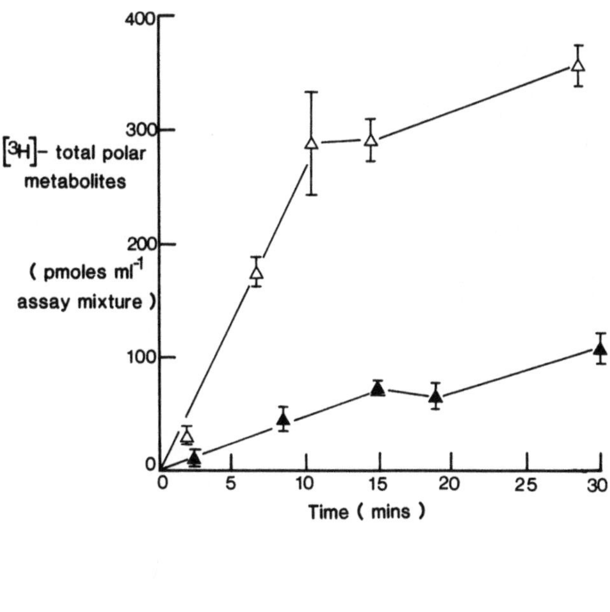

Fig. 4. Benzo [a]pyrene hydroxylase activity of digestive gland microsomes of *M. edulis* showing linearity during the first 15 min of the reaction and with sample size. Microsomes were prepared and assayed at 25 °C as described in Livingstone and Farrar (1984). Values are means and ranges (n = 2); △: 1.4 mg and ▲: 0.28 mg microsomal protein (Lowry assay)

Table 3. Total protein, benzo[a]pyrene hydroxylase and some MFO-associated enzyme activities in digestive gland microsomal fractions of *M. edulis* from some Shetland populations

Population	Total protein mg g^{-1} wet weight	G6PDH	NADH-FERRIED	NADH-CYTCRED	NADPH-CYTCRED	BPH
Gluss Voe	3.11 ± 1.03	11.3 ± 1.7	463 ± 29	12.9 ± 1.2	0.97 ± 0.24	10.5 ± 2.9
Ronas Voe	2.56 ± 0.47	8.2 ± 1.8	641 ± 87	20.6 ± 2.6	1.19 ± 0.22	12.0 ± 8.8
Inner Houb	2.32 ± 0.41	10.6 ± 2.2	764 ± 73[a]	43.1 ± 6.6[b]	2.19 ± 0.85	6.7 ± 1.0
Outer Houb	2.30 ± 0.24	8.4 ± 1.0	706 ± 63[a]	14.1 ± 2.5	1.29 ± 0.22	5.2 ± 1.0

G6PDH, NADH-FERRIRED and NAD(P)H-CYTCRED activities in nmol min^{-1} mg^{-1} protein; BPH in pmol min^{-1} mg^{-1} protein. Values are means ± SEM (n=6; each sample is the pooled digestive glands of three mussels);

[a] Significantly different from Gluss Voe but not from Ronas Voe.

[b] Significantly different from other three populations. Samples collected in July 1982

spectively, the NAD(P)H-cytochrome c and NADH-ferricyanide reductases and microsomal G6PDH (Table 3). In addition, indications have been obtained that the bivalve MFO system is inducible by a number of xenobiotics including hydrocarbons (Anderson 1978; Moore 1980; Moore et al. 1980; Dixon et al. 1984, Guillaume et al. 1983; Payne et al. 1983).

Measurement of the MFO activities of Shetland mussels were first carried out in July 1982 (Table 3) when a reliable assay for BPH activity had been developed. The microsomal yield of total protein was the same in all populations and no differences were seen in G6PDH or NADPH-CYTCRED activities. The individual variability in BPH activity was high and there were no differences between populations or between the pooled data for Ronas Voe and Gluss Voe versus the pooled data for Inner Houb and Outher Houb (>0.1) despite the values for the former two populations appearing higher. The only differences were seen in the NADH-dependent reductase acitvities, which were consistently higher in the Inner Houb population and to some extent also in the Outer Houb population. The interpretation of these differences is virtually impossible at this time, based on the limited knowledge of the mussel MFO system, and the only comment that can be made is that they perhaps indicate a different arrangement of protein components within the MFO system and a different catalytic specificity in the different mussel populations. However, the results for BPH and NADPH-CYTCRED activities would seem to agree with those for blood cell G6PDH and indicate that there has been no specific biochemical response to the presence of hydrocarbons in the Houb of Scatsa environment.

3.3 Conclusion

Mussels exposed in the laboratory to low levels of WAF of North Sea crude oil accumulate aromatic hydrocarbons in their tissues up to concentrations of about 15 to 20 $\mu g\ g^{-1}$ wet weight: furthermore, a linear relationship is observed between the resulting depression of scope for growth or lysosomal latency and the tissue hydrocarbon concentration (Widdows et al. 1982). From these relationships and the observed relatively low levels of hydrocarbons accumulated by the Houb of Scatsa mussels (Table 1), only a very small drop in either scope for growth or lysosomal latency would be predicted, as in fact was observed, and the conclusion made that the degree of stress due to hydrocarbons experienced by the Houb of Scatsa mussels is likely to be slight. The biochemical results appear to be in agreement with this appraisal: no clear indications were obtained of a specific cellular response to hydrocarbons (microsomal BPH and NADPH–CYTCRED and blood cell G6PDH activities) but indications were obtained of a response to a generally hostile environment (adductor muscle PFK and blood total G6PDH activities).

4 The Usefulness of Biochemical Measurements in Biological Effects Monitoring

The results of the Shetlands field study illustrate some of both the positive aspects and the limitations of the current situation of biochemical monitoring. They demonstrate that biochemical differences are discernable between populations of mussels from pristine and polluted or more stressful environments and, as in the case of PFK, that factors such as seasonal variability can be accounted for. However, they also clearly show that the interpretations of the observations are limited and that the biochemical parameters themselves do not measure or identify the extent of any deleterious effect. The final assessment of the mussels' state of health is dependent on the scope for growth information and therefore, in retrospect, this raises the question of the value of such biochemical measurements in biological effects monitoring, and what the emphasis should be for future research.

The importance of studying enzymes such as the MFO stystem, and possibly blood cell G6PDH, which respond to only one type of stressor, remains clear. Placed in a programme of environmental monitoring, including general indices of stress like scope for growth, and chemical analysis of pollutant body burdens, they serve to indicate the possible causative agent in a complex environmental situation. Also, with recent advances in the understanding of marine invertebrate metallothionein and MFO systems, the prospects of making a functional link between pollutant and cellular metabolic consequence are good. In contrast, the usefulness of measurements of activities of PFK and other enzymes of intermediary metabolism is open to question. Although the PFK changes indicate that subtle biochemical responses and possibly adaptive processes are occuring, and aid in the wider understanding of animal/pollutant interactions, they fail in the context of biological effects monitoring because they add nothing of significance to the information available from scope for growth and lysosomal latency measurements. The future usefulness of PFK and similar measurements would seem limited by the unlikelihood of making a simple connection between enzyme activities and animal fitness, a necessary condition for a general stress index (Bayne et al. 1976), and although an index might be developed by a rigorous pursuit of the "range of normality" approach, this could be justified only if it offered significant advantages over existing techniques.

Given such problems, emphasis in future research on biochemical general indices of stress would seem to lie with enzymes where a functional link-up with the deleterious effects of pollutants is more obvious. A recent commendable example of this was the measurement of the release of lysozyme and other protease activity into the haemolymph of mussels following their exposure to copper and caused by a presumed disruption of lysosomal and cellular structure (Pickwell and Steinert 1984). This approach was greatly complemented by the development of an enzyme-linked immunosorbent assay for lysozyme, allowing its measurement quickly and easily in field situations (Steinert and Pickwell 1984) which, as stated in the introduction, is one of the potential strengths of biochemical measurements.

Acknowledgements. The assistance of S. Farrar in measuring benzo[a]pyrene hydroxylase activity is gratefully acknowledged. Drs. B.L. Bayne and M.N. Moore are thanked for critically reading the manuscript. This work forms part of the estuarine and nearshore research programme of the Institute for Marine Environmental Research. It was supported in part by the Department of the Environment (Contract No. DGR/480/47) and by the Shetland Oil Terminal Environment Advisory Group (SOTEAG).

References

Ahmad TA, Chaplin AE (1979) Seasonal variations in the anaerobic metabolism of the mussel *Mytilus edulis* (L.) Comp Biochem Physiol 64B: 351–356

Anderson RS (1978) Benzo[a]pyrene metabolism the American oyster *Crassostrea virginica.* EPA Ecol Res Ser Monogr (EPA-600/3-78-009): 1978

Bayne BL (1973) Physiological changes in *Mytilus edulis* (L.) induced by temperature and nutritive stress. J Mar Biol Assoc UK 53: 39–58

Bayne BL, Anderson J, Engel D, Gilfillan E, Hoss D, Lloyd R, Thurburg FP (1980) Physiological techniques for measuring the biological effects of pollution in the sea. In: McIntyre AD and Pearce JB (eds) Biological effects of marine pollution and the problems of monitoring. Rapp. PV Reun Cons Int Explor Mer 179: 88–99

Bayne BL, Clarke KR, Moore MN (1981). Some practical considerations in the measurement of pollution effects on bivalve molluscs, and some possible ecological consequences. Aquat Toxicol 1: 159–174

Bayne BL, Livingstone DR, Moore MN, Widdows J (1976) A cytochemical and a biochemical index of stress in *Mytilus edulis* (L.) Mar Pollut Bull 7: 221–224

Bayne BL, Moore MN, Widdows J, Livingstone DR, Salkeld P (1979) Measurement of the responses of individuals to environmental stress and pollution: studies with bivalve molluscs. Philos Trans R Soc Lond B Biol Sci 286: 562–581

Bayne BL, Scullard C (1977) Rates of nitrogen excretion by species of *Mytilus* (Bivalvia: Mollusca). J Mar Biol Assoc UK 57: 355–369

Bayne BL, Widdows J, Moore MN, Salkeld P, Worrall CM, Donkin P (1982) Some ecological consequences of the physiological and biochemical effects of petroleum compounds on marine molluscs. Philos Trans R Soc Lond B Biol Sci 297: 219–239

Binder RL, Stegman JJ (1980) Induction of aryl hydrocarbon hydroxylase activity in embryos of an estuarine fish. Biochem Pharmacol 29: 949–951

Blackstock J, Filion-Myklebust C (1983) Environmental and biochemical investigation of some effects of organic pollution in inner oslofjord, Norway. Mar Biol (NY) 73: 155–163

Cairns Jr J, van der Schalie WH (1980). Biological monitoring. Part I – Early warning systems. Water Res 14: 1179–1196

Caldwell RS (1974) Osmotic and ionic regulation in decapod crustacea exposed to methoxychlor. In: Vernberg FJ and Vernberg WB (eds) Pollution and physiology of marine organisms. Academic, New York, pp 197–223

Crabtree B, Newsholme EA (1975) Comparative aspects of fuel utilization and metabolism by muscle. In: Usherwood PNR (ed) Insect muscle. Academic, New York, pp 405–494

De Zwaan A, de Bont AMT, Hemelraad J (1983a) The role of phosphoenolpyruvate carboxykinase in the anaerobic metabolism of the sea mussel *Mytilus edulis* L. J. Comp Physiol 153: 267–274

De Zwaan A, de Bont AMT, Zurberg W, Bayne BL, Livingstone DR (1983b) On the role of strombine formation in the energy metabolism of adductor muscle of a sessile bivalve. J Comp Physiol 149: 557–563

Dixon DR, Jones I.M., Harrison FH (1984) Cytogenetic evidence of inducible processes linked with xenobiotic metabolism in larval and adult *Mytilus edulis.* Mar Environ Res 14: 495–496

Donkin P, Evans SV (1984) Applications of steam distillation to the analysis of petroleum hydrocarbons in water and mussels *(Mytilus edulis)* from dosing experiments with crude oil. Anal Chim Acta 156: 207–219

Ebberink RHM, De Zwaan A (1980) Control of glycolysis in the posterior adductor muscle of the sea mussel *Mytilus edulis*. J Comp Physiol 137: 165–171

Gabbott PA, Cook PA, Whittle MA (1979) Seasonal changes in glycogen synthetase activity in the mantle tissue of the mussel *Mytilus edulis* (L.): regulation by tissue glucose. Biochem Soc Trans 7: 895–896

Gillaume JR, Carles D, Leveau M, Betrand JC, Gilewicz M (to be published 1983) Induction of cytochrome P-450 by hydrocarbons in a mollusc bivalve: *Mytilus galloprovincialis*. In Proceedings of the 5th Conference of the European Society for Comparative Physiology and Biochemistry. Physiological and biochemical aspects of marine biology. Springer, Berlin, Heidelberg New York

Kacser H, Burns JA (1979) Molecular democracy: Who shares the controls. Biochem Soc Trans 7: 1149–1161

Kinter WB, Merkens LS, Janicki RH, Guarino AM (1972) Studies of the mechanism of toxicity of DDT and polychlorinated biphenyls (PCBs): disruption of osmoregulation in marine fish. Environ Health Perspect 1: 169–173

Lee RF, Davies JM, Freeman HC, Ivanovici A, Moore MN, Stegeman J, Uthe JF (1980) Biochemical techniques for monitoring biological effects of pollution in the sea. In: McIntyre AD and Pearce JB (eds) Biological effects of marine pollution and the problems of monitoring. Rapp P V Reun Cons Int Explor Mer 179: 48:55

Livingstone DR (1975) A comparison of the kinectic properties of pyruvate kinase in three populations of *Mytilus edulis* (L.) from different environments. In: Barnes H (ed) Proc 9th Europ Mar Biol symp. The Aberdeen University Press, Aberdeen, pp 151–164

Livingstone DR (1981) Induction of enzymes as a mechanism for the seasonal control of metabolism in marine invertebrates: glucose-6-phosphate dehydrogenases from the mantle and hepatopancreas of the common mussel *Mytilus edulis* (L.). Comp Biochem Physiol 69B: 147–156

Livingstone DR (1982a) General biochemical indicses of sublethal stress. Mar Pollut Bull 13: 261–263

Livingstone DR (1982b) Energy production in the muscle tissues of different kinds of molluscs. In: Addink ADF and Spronk N (eds) Exogenous and endogenous influences on metabolic and neutral control. Pergamon, Oxford, pp 257–274

Livingstone DR, Clarke AR (1983). Seasonal changes in hexokinase from the mantle tissue of the common mussel *Mytlus edulis* (L.) Comp Biochem Physiol 74B: 691–702

Livingstone DR, Farrar SV (to be published 1984) Tissue and subcellular distribution of enzyme activities of mixed-function oxygenase and benzo[a]pyrene metabolism in the commun mussel *Mytilus edulis* (L.). Sci Total Environ

Livingstone DR, Widdows J, Fieth P (1979) Aspects of nitrogen metabilism of the common mussel *Mytilus edulis*: adaptation to abrupt and fluctuating changes in salinity. Mar Biol (NY) 53: 41–55

Lowry OH, Rosebrough NJ, Farr AL, Randall RJ (1951) Protein measurement with the folin phenol reagent. J Biol Chem 193: 265–275

Moore MN (1980) Cytochemical determination of cellular responses to environmental stressors in marine organism. In: McIntyre AD and Pearce JB (eds) Biological effects of marine pollution and the problems of monitoring. Rapp P V Reun Cons Int Explor Mer 179: 7–15

Moore MN (1982) Lysosomes and environmental stress. Mar Pollut Bull 13: 42–43

Moore MN, Livingstone DR, Donkin P, Bayne BL, Widdows J, Lowe DM (1980) Mixed function oxygenase and xenobiotic detoxication/toxication systems in bivalve molluscs. Helgol Wiss Meeresunters 33: 278–291

Newsholme EA, Crabtree B (1973) Metabolic aspects of enzyme activity regulation. Symp Soc Exp Biol 26: 429–460

Newsholme EA, Crabtree B (1981) Flux-generating and regulating steps in metabolic control. Trends Biochem Sci 62: 53–56

Newsholme EA, Start C (1973) Regulation in metabolism. Wiley London, p 349

Payne JF, Rahimtula A, Lim S (1983) Metabolism of petroleum hydrocarbons by marine organisms: the potential of bivalve molluscs. In: 26th Annual Meeting of the Canadian Federation of Biological Societies, vol 26, No PA-235

Pickwell GV, Steinert SA (1984) Serum biochemical and cellular responses to experimental cupric ion challenge in mussels. Mar Environ Res 14: 245–265

Rosenberg DM, Resh VH, Balling SS et al. (1981) Recent trends in environmental impact assesment. Can J Fish Aquat Sci 38: 591–624

Schmidt-Nielsen B, Sheline J, Miller DS, Deldonno M (1977) Effect of methylmercury upon osmoregulation, cellular volume, and ion regulation in winter flounder *Pseudoleuronectes americanus.* In: Vernberg FJ, Calabrese A, Thurberg FP and Vernberg VB (eds) Physiological responses of marine biota to pollutants. Academic, New York, pp 105–117

Steinert SA, Pickwell GV (1984) Quantitative determination of lysozyme in the hemolymph of *Mytilus edulis* by the enzyme-linked immunosorbent assay (ELISA). Mar Environ Res 14: 229–243

Thompson RJ (1972) Feeding and metabolism in the mussel *Mytilus edulis* (L.). Ph. D. thesis, University of Leicester

Weber G (1963) Behaviour and regulation of enzyme systems in normal liver and in hepatomas of different growth rates. In: Weber G (ed) Advances in enzyme regulation. Pergamon, Oxford pp 321–340

Widdows J, Bakke T, Bayne BL et al. (1982) Responses of *Mytilus edulis* on exposure to the water-accomodated fraction of North Nea oil. Mar Biol (NY) 67: 15–31

Widdows J, Bayne BL, Donkin P, Livingstone DR, Low DM, Moore MN, Salkeld PN (1981) Measurement of the responses of mussels to environmental stress and pollution in Sullom Voe: a base-line study. Proc R Soc Edinb Sect B (Biol Sci) 80

Widdows J, Moore SL, Clarke KR, Donkin P (1983) Uptake, tissue distribution and elimination of $1\text{-}^{14}C$ naphtalene in the mussel *Mytilus edulis.* Mar Biol 76: 109–114

Zaba BN (1981) Glycogenolytic pathways in the mantle tissue of *Mytilus edulis* (L.) Mar Biol Lett 2: 64–67

Zaba BN, Gabbott PA, Davies JI (1981) Seasonal changes in the utilization of ^{14}C- and ^{3}H-labelled glucose in a mantle tissue slice preparation of *Mytilus edulis* (L.) Comp Biochem Physiol 70B: 689–695

Zandee DI, Kluytmans JH, Zurburg W, Pieters H (1980) Seasonal variations in biochemical composition of *Mytilus edulis* with reference to energy metabolism and gametogensis. Neth J Sea Res 14: 1–29

Zurburg W, Ebberink RHM (1981). The anaerobic energy demand of *Mytilus edulis.* Organ specific differences in ATP-supplying processes and metabolic routes. Mol Physiol 1: 153–164

Copper and the Mussel *Mytilus edulis* L.

J. DAVENPORT and K.J. REDPATH[1]

1 Introduction

Copper is a beneficial trace element required by all marine organisms, yet is one of the most poisonous of heavy metals when present in excess. Copper is essential for the proper action of several enzymes (e.g., cytocrome oxidase, lysine oxidase and tyrosinase) and appears to function by forming strong complexes with proteins. In some molluscs copper is especially important because it is responsible for the oxygen binding haemolymph pigment haemocyanin. The mussel *Mytilus edulis* L. possesses no such pigment, but copper is still a necessary constituent of *Mytilus* haemolymph proteins. The known deleterious effects of excess copper are also largely associated with proteins; enzyme activities and electron transport reactions may be interfered with, while membrane permeabilities and cell division can also be affected.

The effects of copper upon *Mytilus edulis* have attracted study from two points of view. Firstly, the mussel is an easily identified, long-lived, sessile and cosmopolitan species which accumulates many pollutants and therefore would appear to possess virtually all the attributes of an ideal biological indicator of pollution that were proposed by Butler et al. (1971). The species even flourishes in the brackish waters of estuaries which are often close to concentrations of industry, and so meets the additional indicator criteria proposed by Haug et al. (1974). Because of its apparent suitability for pollutant studies, it has been suggested that *Mytilus* should be used in an economical global marine monitoring exercise (the "Mussel Watch" proposed by Goldberg 1975). Secondly, the mussel is perhaps the most important hard macrofouling organism of static man-made structures, in terms of bulk and increased loading (e.g., oil rigs, buoys, jetties, power station cooling systems) as well as the engine cooling systems of shipping. In some circumstances, especially power stations, chlorine is the main antifouling agent used against mussels, though recent environmental legislation is likely to restrict its use in the USA. However, copper has been used since the 17th century to combat marine fouling (see Woods Hole 1952 for a review of the early literature), firstly as copper sheathing to protect the hulls of ships (mainly against barnacles and seaweed rather than mussels), but latterly as a constituent of antifouling paints, a component of Cu/Ni alloys for marine structures, and as an electrolytically

1 N.E.R.C. Unit of Marine Invertebrate Biology, Marine Science Laboratories, University College of North Wales, Menai Bridge, Gwynedd, North Wales, U.K.

Toxins, Drugs, and Pollutants in Marine Animals
(ed. by Bolis et al.)

generated product of antifouling electrode systems. Workers have been interested, therefore, not only in the subtle sublethal effects of copper upon mussels, but also in determining the minimum concentration of copper which will either prevent the settlement of mussel spat, or at least kill young mussels before they become large enough to present a serious fouling problem.

The present paper summarizes existing information concerning copper and *Mytilus,* and presents new data concerning mussel shell growth at low copper concentrations.

2 Chemistry of Copper in Sea Water

A detailed consideration of the chemistry of copper in sea water is outside the scope of this paper and relevant reviews may be found elsewhere (e.g., Zirino and Yamamoto 1972; Dyrssen and Wedborg 1974; Ahrland 1975; Whitfield 1975; Mantoura et al. 1978; Lewis and Cave 1982). However, in any consideration of the biological literature concerning copper and *Mytilus* it is necessary to be aware of the following:

1. The solubility of copper salts in aerated sea water (pH 7.8 to 8.2; temperature 18 to 23°C) is only 0.4–0.8 ppm (Krauskopf 1956; Bryan 1971). At concentrations higher than this excess copper will be precipitated (primiarily as $CuCO_3$). Some workers (e.g., Brown and Newell 1972) have achieved higher copper concentrations in solution by the use of complexed copper (e.g., as copper sodium citrate), but it has to be said that some workers appear to have been unaware of this low solubility limit and have probably unwittingly exposed test organisms to copper in particulate as well as soluble form.
2. Like other metals (see Burton 1979) copper can exist in sea water systems in a variety of states ranging from free ions, through inorganic/organic complexes, to dispersed colloids and copper adsorbed onto particulate material. Even in solution there are many chemical species of copper (see Figs. 1 and 2), whose proportions are altered by pH, salinity and temperature. It has been tacitly assumed by many that the toxic effects of copper are in the main due to free cupric (II) ions, which make up only about 1–2% of total copper in "normal" sea water, but direct evidence for this is lacking or contradictory (see Lewis and Cave 1982).
3. Unusual environmental conditions may alter the concentration and species composition of copper. In natural waters containing hydrogen sulphide it is probable that copper concentrations will be reduced (Bryan 1971) since sulphide ions are strong ligands, but copper sulphide is insoluble so will be precipitated. Conversely, the addition of chlorine to the sea (mainly in the cooling water of large electricity generating plants) results in alterations in the complexing power of sea water and this causes a release of bound copper and heightened copper toxicity (Carpenter and Smith 1978). Inorganic chelating agents such as EDTA are known to reduce the toxicity of sea water of heightened copper concentration (e.g. Spencer 1957; Erickson et al. 1970). It is feasible that natural chelators (e.g., exudates from algal blooms) may also alter the bioavailability of copper.

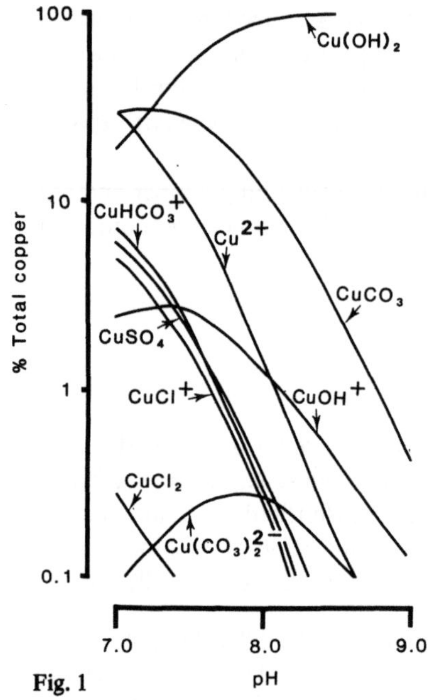

Fig. 1

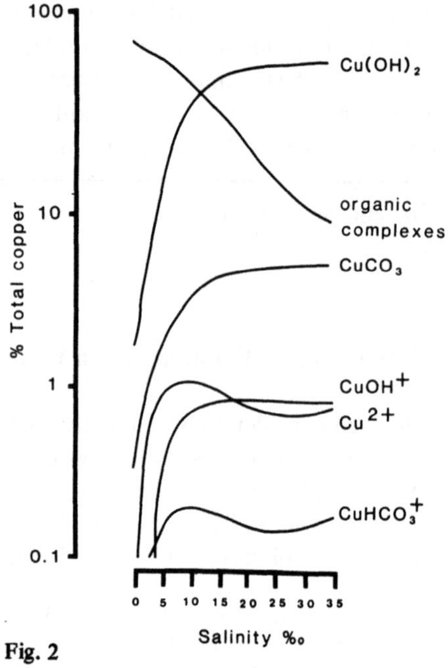

Fig. 2

Fig. 1. Calculated distribution of the chemical species of copper in sea water as a function of pH. (After Zirino and Yamamoto 1972)

Fig. 2. Calculated distribution of the chemical species of copper in sea water as a function of salinity. (After Mantoura et al. 1978)

3 Environmental Copper: Sources and Levels

Copper enters the environment as terrestrial runoff, aeolian input and geothermal addition (see Lewis and Cave 1982). Terrestrial runoff appears to be the major source, so copper concentrations are consenquently highest in nearshore waters. There is some dispute natural levels of copper; Riley and Taylor (1968) reported a level of 5 ppb Cu from coastal waters; while Foster (1976) found a concentration of only 1.4 ppb in the Menai Strait. Since Foster's measurement, a series of records have been taken from sea water from the Menai Strait by a variety of analytical techniques; Davenport and Manley (1978) estimaed 8 ppb Cu with a copper electrode, while Manley (1979) later obtained an atomic absorption spectrophotometric value of 4 ppb. The authors of the present review have recently measured levels of about 2 ppb Cu in sea water analysed with a Radiometer ISS 820 Ion Scanner, so a value of about 2–5 ppb Cu seems reasonable for coastal sea water, and is consistent with the results summarized by Slavin (1980).

Environmental copper input by Man probably stems mainly from mining operations and there is a tendency, in the United Kingdom at least, for copper pollution to be thought of as a problem of the past, when environmental controls were not operating. However, copper may also be added to the sea by power stations, desalination plants and as a by-product of industrial processes including oil refining; dredging operations may also release copper from metalliferous sediments. Unfortunately, water column copper levels in areas affected by Man's industrial activities have rarely been estimated. Although Martin et al. (1975) reported copper levels of 230–520 ppb Cu in the waste discharged by the Los Angeles outfall systems, Lewis and Cave (1982) suggest that values exceeding 30 ppb Cu are probably unusual and very localised, not least because the particulate material of turbid estuarine/coastal water will readily adsorb copper. Sedimentary copper levels can of course be several orders of magnitude greater. For example: Bryan (1971) reported a concentration of 407 ppm Cu in dried sediments from the Tamar estuary which is contaminated by the effluent from old mine workings in its watershed.

4 Toxic Concentrations

Unfortunately, many of the published heavy metal toxicity data are compromised to a greater or lesser extent by the use of static experimental techniques, that is the placement of organisms in vessels containing test medium which is not continually replenished. Consider a single mussel placed in a 1-l vessel filled with sea water containing 0.1 ppm copper; let us assume that the test medium is replaced daily. The maximum amount of copper that the mussel can be exposed to or take up in a 24-h period is only 0.1 mg; the animal will rapdily change the effective concentration of the test medium by its own powers of concentration. On the other hand, a mussel ventilating its mantle cavity at 50 ml min^{-1} and held in a 1-l vessel supplied continually at a high flow rate (say 500 ml min^{-1}) with sea water containing 0.1 ppm Cu, can be exposed to 7.2 mg of copper passing through its mantle cavity each day, and will be quite unable to change the test medium concentration to a significant extent. Clearly the latter dynamic experiment is far more representative of the real world and yields more objective data because the animals tested cannot control the metal level of their surroundings; it also permits accurate comparison of the responses of animals of differing size (and accumulating power).

The variation in experimental approach and differences in criteria of toxicity have led to a considerable variety in published lethal concentrations for *Mytilus edulis*, but there has been a general tendency for the more recent values to be lower – mainly because through flow systems have been used. Marks (1938) reported that the toxicity threshold for *Mytilus* was about 200 ppb; Scott and Major (1972) confirmed this value in their static experiments, finding that 200 ppb caused 100% mortality in about 12 days, while most animals survived 100 ppb. In contrast, Davenport and Manley (1978) found that 90 ppb Cu killed 50% of animals in about 10 days in their through flow studies and Manley (1980) went on to show that even 15 ppb added copper (ca. 20 ppb Cu in toto) killed 50% of animals after about 1 month. This is consistent with the findings of Martin (1979) and the earlier work of Adema et al. (1972), who worked with a through flow system but with artificial sea water.

It is well known that different stages of the life history of organisms exhibit different sensitivities to pollutants, and it has often been suggested that larval stages are less tolerant than adults. The mussel reproduces by broadcasting gametes and the fertilized eggs give rise to a short lived planktonic trochophore stage (ca. 24 h) itself succeeded by the veliger phase which remains in the plankton for several weeks. Settlement is a quite complex procedure in *Mytilus,* with pediveligers first settling on weed and later drifting to their final resting place by using their byssus threads as drogues. There appear to be no published data concerning the effects of heightened copper on the viability of *Mytilus* gametes or upon young larvae, although Beaumont and Tserpcs (pers commun) have recently found that exposure of veliger and pediveliger stages to 200 ppb Cu fails to cause significant mortality in 2 weeks. Settlement stages have attracted some attention; Wisely and Blick (1967) exposed *Mytilus* pediveligers/plantigrades to sea water solutions of copper citrate. Unfortunately their use of citrate, plus an unusual toxicity criterion, makes comparison with results for adult mussels virtually impossible; they found that 50% of mussel larvae were killed by c. 22 ppm Cu in 2 h. More recently Sheffrin (1982) has shown that the settling stages of *Mytilus* do indeed appear to be more tolerant of added copper than adults; 137 ppb Cu required 52–53 days to kill 50% of *Mytilus* plantigrades (both Sheffrin's concentrations and those of Beaumont and Tserpes were checked by potentiometric stripping analysis).

5 Accumulation, Storage and Excretion of Copper

Natural levels of copper in the tissues of mussels living in uncontaminated areas are close to 10 mg Cu kg^{-1} (dry weight) (Delhaye and Cornet 1975; Simkiss et al. 1982). Mussel tissues contain about 70% water, so if an environmental level of 2–5 ppb Cu is assumed, then a volume of mussel tissue will contain about 10^3 times the amount of copper contained in an equivalent volume of sea water. This concentrating power is impressive, but is in fact considerably less than the mussel's ability to concentrate Fe, Cd or Zn (10^4 to 10^5 fold). Boyden (1977) found that there was a size effect on copper concentration in *Mytilus.* He found that:

$$\text{if } Y = aX^b \qquad (1)$$

where Y = total copper content/individual mussel
and X = weight of individual

then for the mussel b = 0.8, indicating that smaller (and therefore younger) mussels contain higher concentrations than larger animals.

When mussels are exposed to heightened copper concentrations, the tissue levels also rise rapidly. Scott and Major (1972) exposed mussels to 300 ppb Cu in a static system and found that tissue copper levels rose from about 10 mg Cu kg^{-1} (dry weight) to about 25 mg Cu kg^{-1} (dry weight) after 24 h (see Fig. 3); in the subsequent 24-h period there was a decline back to the initial tissue concentration. Such a decrease was not confirmed by Dehlaye and Cornet (1975), but they worked at rather higher concentrations. Scott and Major's results appear to indicate that their mussels had removed all of the copper from the static test solution surrounding them within 24 h, and that the copper was subsequently excreted, presumably in particulate form. Even though they were apparently able to excrete copper, the mussels showed 50%

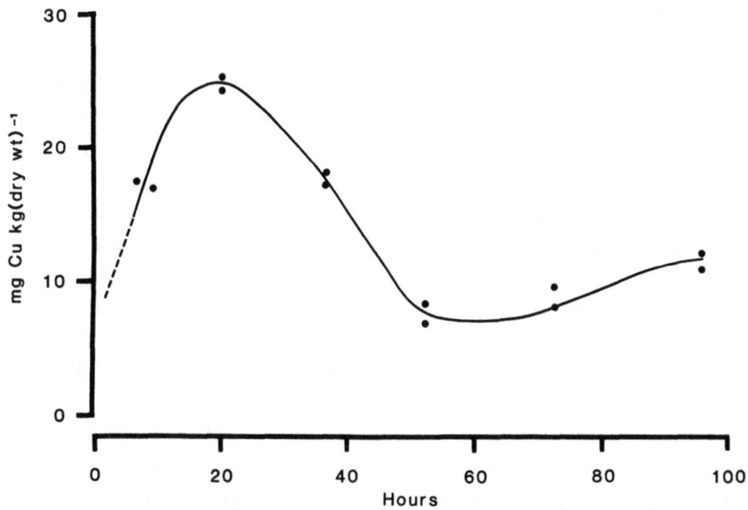

Fig. 3. Accumulation and excretion of copper by mussels exposed to 300 ppb Cu at time 0. (After Scott and Major 1972)

mortality after about 6 days, so the transient presence of copper had clearly caused irreversible damage.

Only one study of copper accumulation by mussels transported to a copper-contaminated field site appears to have been performed (although this was implicit in Goldberg's "Mussel Watch" proposals); Boyden (1977) investigated mussels transferred from the "clean" Helford estuary to the nearby contaminated Restronguet Creek in Cornwall, U.K. Despite spending 5 months in their new surroundings, it appears that larger mussels at least did not equilibrate completely with the environment, since the b value obtained for Eq. (1) was only 0.61. This result, together with Scott and Major's finding that copper is rapidly excreted, tends to suggest that *Mytilus* is a poor indicator species as far as copper is concerned. This conclusion was also reached by Phillips (1976a, 1976b), who found that copper concentrations in mussels varied with season, mainly because of wet weight changes caused by the build up and loss of gonadal material. He also found erratic variations in copper level associated with temperature, salinity and the presence of other metals. Phillips also pointed out that, despite evidence of large differences between cadmium and lead concentrations in mussels collected from industrialized and non-industrialized areas (40-fold for Cd; 14-fold for Pb), there were usually no more than 4-fold differences in copper concentration. Phillips hypothesised that there might be specific copper transporting sites (on the gills?) which have evolved because of the biological importance of copper to *Mytilus*. On the other hand, Davies and Pirie (1980), who performed a "Mussel Watch" project in Scottish coastal waters did find higher copper levels in the industrialized areas of the Forth, Clyde and Tay estuaries (values were roughly double those found in open coast areas). Even in this case a note of caution should perhaps be struck; low salinities are known to alter copper speciation (see Fig. 2), while Phillips (1977, 1978) reported higher metal concentrations from the low salinity Baltic Sea. It might be expected, therefore, that estuarine mussels should contain more copper than coastal animals, even in the absence of industrial inputs. Young and Mc Dermott (1975), Young and Alexander (1977) and Young et al. (1979) certainly demonstrated heightened tissue copper levels in mussels living in Californian harbours contaminated by municipal waste water and antifouling paints; Young and Alexander found that the digestive glands of mussels living in an enclosed harbour contained 127 mg Cu kg^{-1} dry tissue compared with 16 mg Cu kg^{-1} (dry weight) in coastal control animals – and eight-fold difference. In harbours which were better flushed by tidal ebb and flow the results were less dramatic, with "contaminated" mussels having a concentration of only 30 mg Cu kg^{-1} (dry weight) (double the control value). These studies show that *Mytilus* may indeed give warning of environmental contamination by copper, but it is not at all clear that mussels give a proper integration of environmental copper levels/changes in the manner required of an ideal indicator species.

The storage of accumulated copper by *Mytilus* is a somewhat obscure phenomenon too. Several metals are concentrated in particular body tissues, especially the kidney (see Table 1), but copper appears to be generally distributed throughout the body. Iron and zinc are stored, as in many other invertebrates, in the form of insoluble (and therefore presumably innocuous) metal-containing granules (see Lewis and Cave 1982 for review). However, although George and Pirie (1980) did find a little copper in granules from the kidney of *Mytilus*, the amount was small and partially derived from the electron probe instrument employed in their analysis.

In vertebrates (see Kagi and Nordberg 1979) it is common to find specific low molecular weight, cysteine-rich proteins (metallothioneins) which bind metals (Zn/Cd/Hg/Cu) and are believed to detoxify them, presumably by competition with vulnerable proteins such as enzymes (see Roesijadi 1981 for a review of the significance of metallothioneins in marine invertebrates). Noël-Lambot (1976) identified cadmium binding proteins similar to metallothioneins in *Mytilus*, but found that zinc and copper were associated with high molecular weight proteins. This appears not to be the whole story however, since Talbot and Magee (1978) isolated low molecular weight copper binding proteins from mussels living in a cadmium polluted area near Melbourne, Australia. These workers suggested that the proteins were metallothionein like, and that their synthesis might be stimulated by the presence of excess environmental cadmium. Viarengo et al. (1980) have recently found that the closely

Table 1. Tissue concentrations of heavy metals in ***Mytilus edulis*** (in mg metal kg^{-1} dry weight.
(After George and Pirie 1980)

Tissue	Cd	Cu	Fe	Pb	Zn
Muscle + foot	1.6	4.8	230	1.4	60
Mantle	2.2	8.8	400	2.0	61
Gills + palps	2.0	13.8	530	1.2	72
Kidney	56.0	25.7	2260	98.0	995
Viscera + rest	8.8	14.6	640	2.9	110

related mussel, *Mytilus galloprovincialis* is capable of synthesising a copper-binding, metallothionein-like low molecular weight protein in response to the presence of heightened copper alone. They found that the protein was not present in detectable concentrations in control mussels, or in mussels exposed to copper for only 4 h, but appeared after 24–48 h when mussels were exposed to 80 ppb Cu. Tentatively it may be suggested that copper accumulated from natural copper concentrations in sea water is stored bound to high molecular weight proteins, but that mussels have the ability to synthesise metallothionein (or a protein with similar attributes) when faced with an excess copper burden.

Excretion of copper by *Mytilus* has been relatively poorly studied. The data of Scott and Major (1972) indicated that copper was excreted in non soluble form and the authors of the study suggested that copper was excreted in mucus and/or faecal material (they had noted that the presence of 300 ppb Cu stimulated copious mucus production). No confirmation of these excretory pathways appears to have been published. George et al. (1976), and Coombs and Keller (1981) provide good evidence that the proteinaceous byssal threads and discs of *Mytilus* concentrate a wide range of heavy metals and provide an effective excretion route. However, it is well known that the mussel *reduces* the rate of byssus production when under stress, and two studies (Martin et al. 1975; Davenport 1977) have indicated that mussels exposed to heightened copper concentrations secrete fewer byssal threads. It is probable therefore, that the byssus gland can only function as an excretory organ for copper at relatively low external concentrations of Cu.

6 Behavioural, Physiological, and Metabolic Effects

Several workers have noted that larval and adult *Mytilus* close their shell valves to isolate their tissues from the environment when they are exposed to high copper concentrations (Wisely and Blick 1967; Scott and Major 1972; Delhaye and Cornet 1975; Davenport 1977), a response not elicited by other heavy metals (Manley 1979). Davenport (1977) showed that this response could allow mussels to survive intermittent exposure to lethal copper concentrations (500 ppb), provided that animals were able to gain regular access to clean sea water. He pointed out that this was another reason for doubting the efficacy of *Mytilus* as a biological indicator of copper, since it was feasible for transient or intermittent high environmental copper levels not to be reflected in raised tissue concentrations. Subsequent studies (Davenport and Manley 1978; Manley and Davenport 1979; Manley 1983) have demonstrated that the full shell valve closure response to copper is unlikely ever to be deployed in the field because a concentration of 200 pb Cu is needed to trigger it. On the other hand, a disturbance in normal shell valve movements is elicited by a concentration of only 20 ppb added Cu (Davenport and Manley 1978) while Manley (1983) recorded sustained decreases in ventilatory activity (reflected in reduced mantle cavity oxygen tensions) in the presence of 25 ppb added copper; both of these concentrations are apparently approached or exceeded in areas affected by industry (see Lewis and Cave 1982).

The effects of copper upon oxygen consumption, filtration rate and heart rate have attracted the attention of several scientists, but the study of all three parameters is complicated by the ability of *Mytilus* not only to close the shell values, but also to alter pumping rates without changing shell gape by control of the exhalant siphon (see Davenport 1979a, 1979b; Davenport and Woolmington 1982). Scott and Major (1972) were apparently not aware of this subtlety, and simply waited until animals were showing a shell gape before making oxygen uptake measurements. They demonstrated pronounced depression of oxygen uptake, but since almost all of their concentrations were high enough to trigger exhalant siphon closure, and consequently greatly reduce mantle cavity irrigation, the actual values are of dubious validity. Brown and Newell (1972) and Manley (1983) prepared animals for respirometric measurements by cutting the posterior shell adductor muscles, thus ensuring continuity between the mantle fluid and external medium; Delhaye and Cornet (1975) ground away the edges of the shell valves to achieve the same aim. All three studies revealed reductions in oxygen consumption caused by the presence of copper, but in Brown and Newell's study, the use of copper citrate at a grossly unphysiological concentration (500 ppm) makes interpretation difficult. Delhaye and Cornet obtained similar data to Manley; a reduction of about 50% in oxygen uptake was seen in response to 200–250 ppb Cu. At lower copper concentrations there were undoubtedly lesser depressive effects, but the considerable natural variability of oxygen uptake in *Mytilus* masks the effects of the heavy metal.

Filtration rate and respiration are intimately associated in bivalve molluscs, so it is not surprising to find that added copper suppresses filtration as well as oxygen uptake in *Mytilus*. Abel (1976) recorded a reduction of 50% in filtration rate at copper concentrations (static system) between 80 and 230 ppb Cu, which agrees well with the respiration rate effects described above. In his through flow experiments Manley (1983) recorded a 50% depression of filtration rate at a rather lower copper concentration (30 ppb added Cu). He was also able to discern a threshold for disturbance of filtration rate at about 10 ppb added Cu.

There is some disagreement about the mode of interference of copper with filtration and respiration. Brown and Newell (1972) concluded that the reduction in respiration of *Mytilus* in the presence of copper citrate at a concentration of 500 ppm was due to the suppression of ciliary action rather than a direct inhibition of cellular respiratory activity. However, Manley (1983) reported that the frontal gill cilia of *Mytilus* (which are admittedly responsible for food transport rather than driving the respiratory/filtration current) were unaffected by 500 ppb Cu, but that the epithelial tissues of the ctenidia broke down at this concentration so that the gill filaments became "unzipped" thus rendering any action of the lateral cilia (responsible for ventilation) ineffective.

Changes in few other physiological functions have been studied in relation to copper. Reduced byssus thread production has already been mentioned (in the preceding section), while Scott and Major (1972) monitored heart rate changes in response to elevated copper concentrations. Although these workers reported significant suppression of heart rate at concentrations above 400 ppb Cu, it is difficult to avoid the conclusion that the animals were simply isolating themselves from their surroundings. in the manner described by Davenport (1977), and showing the normal bradycardia associated with such isolation.

Far more interesting and environmentally relevant is the recent finding by Viarengo et al. (1981) that 80 ppb added Cu has a profound effect upon protein metabolism in *Mytilus galloprovincialis.* They found that uptake of amino acids by gills, digestive gland and mantle was reduced by 90–95% in 7 days; protein synthesis was depressed by up to half while ATP concentrations fell by 25–30% in the same period. Viarengo et al. employed a static experimental approach with the test medium changed daily; since each 4–6 cm mussel was placed in only 1 l of sea water it would seem likely that similar deleterious metabolic effects would be elicited by appreciably lower copper concentrations in a through flow system. Recent electrophoretic studies stress that *Mytilus galloprincialis* is very closely related to, if not synonymous with *Mytilus edulis* (Burfit pers commun), so it is almost certain that these data are equally applicable to the latter species.

7 Growth

Growth results from the concerted action of a range of behavioural, physiological and biochemical functions, and might therefore be expected to be especially sensitive to the action of pollutants. Until relatively recently it would have been difficult to carry out growth studies upon *Mytilus* because of the necessary investment of time and material. However, Strömgren (1975) introduced the laser diffraction technique which allows shell growth measurements to be made upon small mussels on a daily

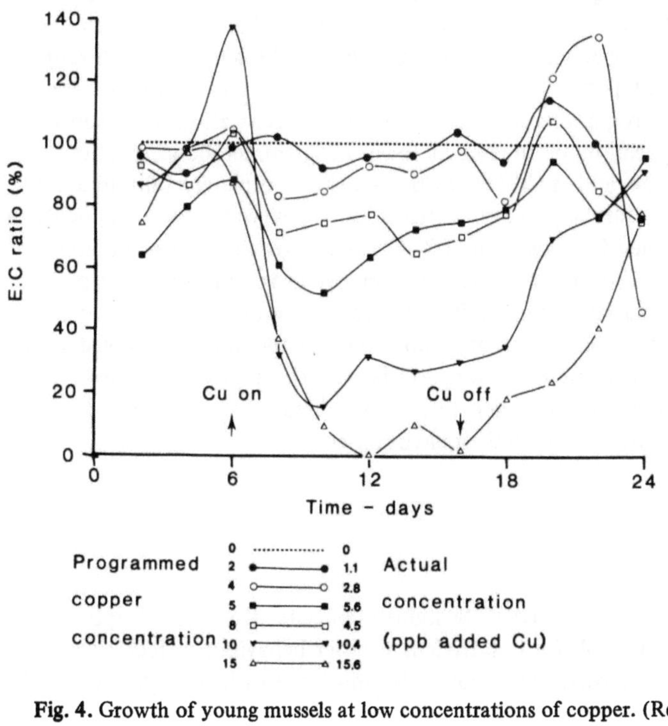

Fig. 4. Growth of young mussels at low concentrations of copper. (Redpath, in press)

basis, thus permitting much shorter experiments. Using this technique Manley (1979) reported that shell growth was significantly depressed at a concentration of 10 ppb added Cu, and that growth inhibition became greater as exposure to this concentration continued. Strömgren (1982) has also recently published a description of the effects of heavy metals upon *Mytilus*. These confirm depression of shell growth at low copper concentrations (in the 5–20 ppb Cu range), but there are puzzling features in Strömgren's paper which make comparison difficult. The sea water used in his investigation was taken from a source affected by an ore processing plant and was said to have a "relatively high background content", yet the concentration quoted (1.0 ppb Cu) is in fact lower than all published values for coastal sea water! Secondly, and more importantly, Strömgren stated that copper concentrations above 5 ppb Cu caused "apparently permanent closing of the shell valves" so that reduced growth was partly caused by "starvation stress". This observation is quite inconsistent with the findings of all other workers who have recorded shell valve closure in response to copper; at least 200 ppb appears to be required to trigger complete isolation, and such isolation does not persist beyond 15–24 h (Scott and Major 1972; Davenport 1977). Finally, the temperature of the system fluctuated greatly, presumably with attendant effects on copper speciation and stress upon the test animals. Strömgren reported that shell growth in mussels exposed to 10 ppb Cu for 3 weeks recovered rapidly after they were returned to clean sea water; this finding is similar to that reported by Manley (1979). Recently, one of the authors of the present review (K.J.R.) has extended the observations of Manley upon Menai Strait *Mytilus* by recording growth rates at a range of low copper concentrations whilst checking (with potentiometric stripping analysis) the actual experimental copper levels achieved against the nominal ones determined by the relationship between sea water and copper solution flow rates produced by peristaltic pumps; the temperature throughout was held at a constant 14°C. The results are displaved in Fig. 4. It may be seen that agreement between programmed and actual copper concentrations is reasonable, though discrepancies appear to be greater at lower programmed concentrations. A concentration of 15.6 ppb added Cu (ca. 17.6 total Cu) completely inhibited shell growth, but the affected animals recovered when clean sea water was supplied to them again. 10.4 ppb added Cu caused an 80% decrease in growth rate and there was little sign of recovery until after the stressor was removed. At concentrations of 2.8, 4.5 and 5.6 ppb added Cu there were initial decreases in growth rate (which were more severe the greater the concentration of copper), but the animals showed signs of recovery of shell growth while they were still being exposed to heightened metal concentrations. Only the animals exposed to 1.1 ppb added Cu showed no change in growth rate when copper was added to their environment.

8 Conclusions

Copper is clearly very toxic to mussels. Sustained concentrations of 20 ppb Cu or more kill *Mytilus*, while 10 ppb Cu is sufficient to alter behaviour, affect physiological processes and inhibit growth. Even concentrations of as little as 2–5 ppb added

Cu produce measureable growth checks. These values are less than an order of magnitude higher than copper concentrations prevailing in coastal waters, and lower than levels reported from areas affected by Man's activities.

Such sensitivity has been reported for other organisms; Bougis (1965), for example, reported that growth was inhibited in larvae of the sea urchin *Paracentrotus lividus* when they were exposed to 10 ppb Cu. However, it must be stressed that the sensitivity of *Mytilus* was only revealed when through flow experimental systems were used. Earlier static tests suggested a ten-fold greater tolerance of copper. Experiments with accumulating poisons like heavy metals (or hydrocarbons, pesticides etc.) cannot be treated in the manner of simple salinity or temperature studies and through flow experiments should be regarded as essential in future studies upon *Mytilus*.

References

Abel PD (1976) Effects of some pollutants on the filtration rate of *Mytilus*. Mar Pollut Bull 7 (12): 228–231

Adema DMM, Swaaf Mooy SI, Bais P (1972) Laboratorium-onderzoek over de involoed van koper mossolen *(Mytilus edulis)*. TNO-Nieuws 27: 482–486

Ahrland S (1975) Metal complexes in sea water. In: Goldberg ED (ed) The nature of sea water. Dahlem Konferenzen, Berlin, pp 219–244

Bougis P (1965) Effet du cuivre sur la croissance du pluteus d'Oursin *(Paracentrotus lividus)*. C R Heba Seances Acad Sci 260: 2929–2931

Boyden CD (1977) Effect of size upon metal content of shellfish. J Mar Biol Assoc UK 57: 675–714

Brown BE, Newell RC (1972) The effect of copper and zinc on the metabolism of the mussel *Mytilus edulis*. Mar Biol (NY) 16: 108–118

Bryan GW (1971) The effects of heavy metals (other than mercury) on marine and estuarine organisms. Proc R Soc Lond Biol Sci 177: 389–410

Burton JD (1979) Physioc-chemical limitations in experimental investigations. Philos Trans R Soc Lond Biol Sci 286: 443–456

Butler PA, Andrew L, Bonde GJ, Jernelov A, Reish DJ (1971) Monitoring organisms. In: Ruivo M (ed) FAO Technical conference on marine pollution and its effects on living resources and fishing Rome, 1970. Supplement 1: Methods of detection, measurement and monitoring of pollutants in the marine environment. Fishing News (Books), London, pp 101–112

Carpenter J, Smith CA (1978) Reactions in chlorinated sea water. In: Jolley RL, Gorchov H, Hamilton DH Jr (eds) Environmental impact and health effects, vol 2. Proc 2nd Conf., Gatlinburg, Tenn, USA, pp 195–207

Coombs TL, Keller PJ (1981) *Mytilus* byssal threads as an environmental marker for metals. Aquat Toxicol 1: 291–300

Davenport J (1977) A study of the effects of copper applied continuously and discontinuously to specimens of *Mytilus edulis* (L.) exposed to steady and fluctuating salinity levels. J Mar Biol Assoc UK 57: 63–74

Davenport J (1979a) The isolation response of mussels (*Mytilus edulis* L.) exposed to falling sea water concentrations. J Mar Biol Assoc UK 59: 124–132

Davenport J (1979b) Is *Mytilus edulis* a short term osmoregulator? Comp Biochem Physiol A: 64: 91–95

Davenport J, Manley AR (1978) The detection of heightened sea water copper concentrations by the mussel *Mytilus edulis*. J Mar Biol Assoc UK 58: 843–850

Davenport J, Woolmington AD (1982) A new method of monitoring ventilatory activity in mussels and its use in a study of the ventilatory patterns of *Mytilus edulis* (L.) J Exp Mar Biol Ecol 62: 55–67

Davies IM, Pirie JM (1980) Evaluation of a "Mussel Watch" project for heavy metals in Scottish coastal waters. Mar Biol (NY) 57: 87–93

Delhaye W, Cornet D (1975) Contribution to the study of the effect of copper on *Mytilus edulis* during reproductive period. Comp Biochem Physiol A 50: 511–518

Dyrssen D, Wedborg M (1974) Equilibrium calculations of the speciation of elements in sea water. In: Goldberg ED (ed) The Sea, Vol 5. Marine chemistry. Wiley Interscience, New York, pp 181–195

Erickson SF, Maloney TE, Gentile JH (1970) Effect of nitrilo-triacetic acid on the growth and metabolism of estuarine phytoplankton. J Water Pollut Control Fed 42: 329–335

Foster P (1976) Concentrations and concentration factors of heavy metals in brown algae. Environ Pollut 10: 45–53

George SG, Pirie BJS (1980) Metabolism of zinc in the mussel *Mytilus edulis* (L.): a combined ultrastructural and biochemical study. J Mar Biol Assoc UK 60: 575–590

George SG, Pirie BJS, Coombs TL (1976) Kinetics of accumulation and excretion of ferric hydroxide in *Mytilus edulis* (L.) and its distribution in the tissues. J Exp Mar Biol Ecol 23: 71–84

Goldberg ED (1975) The mussel watch – a first step in global monitoring. Mar Pollut Bull 6(7): 111

Haug A, Melsom S, Omong S (1974) Estimation of heavy metal pollution in two Norwegian fjord areas by analysis of the brown alga *Ascophyllum nodosum.* Environ Pollut 7: 173–193

Kagi JHR, Nordberg M (1979) Metallothionein. Proceedings of the 1st International Meeting on Metallothionein and other Low Molecular Weight Metal-Binding Proteins, Birkhauser, Basel, 378 pp.

Krauskopf KB (1956) Factors controlling the concentrations of thirteen rare metals in sea water. Geochim Cosmochim Acta 9: 1–32B

Lewis AG, Cave WR (1982) The biological importance of copper in oceans and estuaries. Oceanogr Mar Biol Annu Rev 20: 471–695

Manley AR (1980) The responses of marine bivalves to selected heavy metals. PhD Thesis, University of Wales

Manley AR (1983) The effects of copper on the behaviour, respiration, filtration and ventilation activity of *Mytilus edulis.* J Mar Biol Assoc UK 63: 205–222

Manley AR, Davenport J (1979) Behavioural responses of some marine bivalves to heightened sea water copper concentrations. Bull Environ Contam Toxicol 22: 739–744

Mantoura RFC, Dickson A, Riley JP (1978) The complexation of metals with humic materials in natural waters. Estuarine Coastal Mar Sci 6: 387–408

Marks GW (1938) The copper content and copper tolerance of some species of molluscs of the southern California coast. Biol Bull (Woods Hole) 75 (2): 224–237

Martin JM, Piltz FM, Reish DJ (1975) Studies on the *Mytilus edulis* community in Alamitos Bay, California. V. The effects of heavy metals on byssal thread production. Veliger 18: 183–188

Martin JLM (1979) Schema of lethal action of copper on mussels. Bull Environ Contam Toxicol 21: 808–814

Noël-Lambot F (1976) Distribution of cadmium, zinc and copper in the mussel *Mytilus edulis.* Existence of cadmium binding proteins similar to metallothioneins. Experientia (Basel) 32 (3): 324–325

Phillips DJH (1976a) The common mussel *Mytilus edulis* as an indicator of pollution by zinc, cadmium, lead and copper. I. Effects of environmental variables on uptake of metals. Mar Biol (NY) 38: 59–69

Phillips DJH (1976b) The common mussel *Mytilus edulis* as an indicator of pollution by zinc, cadmium, lead and copper. II. Relationship of metals in the mussel to those discharged by industry. Mar Biol (NY) 38: 71–81

Phillips DJH (1977) The common mussel *Mytilus edulis* as an indicator of trace metals in Scandinavian waters. I. Zinc and cadmium. Mar Biol (NY) 43: 283–292

Phillips DJH (1978) The common mussel *Mytilus edulis* as an indicator of trace metals in Scandinavian waters. II. Lead, iron and manganese. Mar Biol (NY) 46: 147–156

Redpath KJ (in press) Growth inhibition and recovery in mussels *(Mytilus edulis)* exposed to copper concentrations. J Mar Biol Assoc UK

Riley JP, Taylor D (1968) Chelating resins for the concentration of trace elements from sea water and their use in conjunction with atomic absorption spectrophotometry. Anal Chim Acta 40: 479–485

Roesijadi G (1981) The significance of low molecular weight, metallothionein-like proteins in marine invertebrates: current status. Mar Environ Res 4: 167–179

Scott DM, Major CW (1972) The effect of copper (II) on survival, respiration and heart rate in the common blue mussel *Mytilus edulis*. Biol Bull (Woods Hole) 143: 679–688

Sheffrin NMH (1982) The effect of impaired water quality on the behaviour of *Mytilus edulis* plantigrades. PhD Thesis, University of Sheffield

Simkiss K, Taylor M, Mason AZ (1982) Metal detoxification and bioaccumulation in molluscs. Mar Biol Lett 3: 187–201

Slavin W (1980) The determination of trace metals in sea water. At Spectrosc 1 (39): 66–71

Spencer CP (1957) Utilization of trace elements by marine unicellular algae. J Gen Microbiol 16: 282–285

Strömgen T (1975) Linear measurements of growth of shells using laser diffraction. Limnol Oceanogr 20: 845–848

Strömgen T (1982) Effects of heavy metals (Zn, Hg, Cu, Cd, Pb, Ni) on the length growth of *Mytilus edulis*. Mar Biol (NY) 72: 69–72

Talbot V, Magee RJ (1978) Naturally occurring heavy metal binding proteins in invertebrates. Arch Environ Contam Toxicol 7 (1): 73–81

Viarengo A, Pertica M, Mancinelli G, Capelli R, Orunesu M (1981) Effects of copper on the uptake of amino acids, on protein synthesis and on ATP content in different tissues of *Mytilus galloprovincialis* (Lam.). Mar Environ Res 4: 145–152

Viarengo A, Pertica M, Mancinelli G, Zanicchi G, Orunesu M (1980) Rapid induction of copper-binding proteins in the gills of metal exposed mussels. Comp Biochem Physiol C 67: 215–218

Whitfield M (1975) The electro-analytical chemistry of sea water. In: Riley JP and Skirrow G (eds) Chemical oceanography, vol 4. Academic, London, p 1–154

Wisely B, Blick RAP (1967) Mortaliy of marine invertebrate larvae in mercury, copper and zinc solutions. Aust J Mar Freshwater Res 18: 63–72

Woods Hole (1952) Marine fouling and its prevention. United States Naval Institute Annapolis Maryland

Young DR, Alexander GV (1977) Metals in mussels from harbours and outfall areas. Annu Rep South Calif Coastal Water Res Proj 159–165

Young DR, McDermott DJ (1975) Trace metals in harbor mussels. Annu Rep South Calif Coastal Water Res Proj 139–142

Young DR, Alexander GV, McDermott-Ehrlich D (1979) Vessel related contamination of southern California harbours by copper and metals. Mar Pollut Bull 10: 50–56

Zirino A, Yamamoto S (1972) A pH-dependent model for the chemical speciation of copper, zinc, cadmium and lead in sea water. Limnol Oceanogr 17: 661–671

Subject Index

Lecture Notes on Coastal and Estuarine Studies

Managing Editors:
R. T. Barber, M. T. Bowman,
C. N. K. Mooers, B. Zeitzschel

Volume 1

Mathematical Modelling of Estuarine Physics

Proceedings of an International Symposium Held at the German Hydrographic Institute, Hamburg, August 24–26, 1978
Editors: **J. Sündermann, K.-P. Holz**
1980. 119 figures, 1 table. VIII, 265 pages. ISBN 3-540-09750-3

Contents: Basic formulations and algorithms. – Tides and storm surges. – Baroclinic motions and transport processes.

Volume 2
D P. Finn

Managing the Ocean Resources of the United States: The Role of the Federal Marine Sanctuaries Program

1982. IX, 193 pages. ISBN 3-540-11583-8

Contents: Introduction. – Case Studies. – Interagency Coordination for the Management of Marine Resources. – The Marine Sanctuaries Program. – The Role of Designating Marine Areas for Special Management. – Recommendations and Conclusions. – Notes. – Alphabetical List of Major References.

Volume 3

Synthesis and Modelling of Intermittent Estuaries

A Case Study from Planning to Evaluation

Editors: **W. R. Cuff, M. Tomczak, Jr.**
1983. VIII, 302 pages. ISBN 3-540-12681-3

The Port Hacking Estuary Program set out in 1973 is analyzed as a case study of model-guided, multidisciplinary ecosystem studies. Organizational difficulties are discussed with the aim of helping future studies of this genre avoid similar mistakes. The book illustrates how a data set that is incomplete from the viewpoint of constructing a predictive dynamic ecosystem model can still be synthesized and used.

Volume 4
H. R. Gordon, A. Y. Morel

Remote Assessment of Ocean Color for Interpretation of Satellite Visible Imagery

A Review

1983. V, 144 pages. ISBN 3-540-90923-0

Contents: Introduction. – The Physics of Ocean Color Remote Sensing: Irradiance Ratio and Upwelling Subsurface Radiance, Atmospheric Effects. – In-Water Algorithms: The Phytoplankton Pigment Algorithms. The 'K' Algorithms. The Seston Algorithms. The Analytic Algorithm. Relationship Between the Algorithms. – Atmospheric Correction. – Application of the Algorithms to CZCS Imagery. – Summary and Conclusions. – Appendix I: The Coastal Zone Color Scanner (CZCS). – Appendix II: Recent Developments: Clear Water Radiance Concept. Accuracy of Pigment Estimates. Applications. – References.

Volume 5
D. C. L. Lam, C. R. Murthy, R. B. Simpson

Effluent Transport and Diffusion Models for the Coastal Zone

1984. IX, 168 pages. ISBN 3-540-90928-1

Contents: Introduction. – Parameterization of Advection and Diffusion Processes. – Mathematical Models with Analytical Solutions. – Marching Technique Solutions for Straight Plume Equations: Effects of Scale Dependent Diffusivity. – Fully Two-Dimensional Computational Technique for Steady Plume Modelling. – Testing Finite Element Plume Using Examples with Analytical Solutions. – Vertification and Application. – References. – Appendix I: Notations. – Appendix II: Subject Index.

Springer-Verlag
Berlin
Heidelberg
New York
Tokyo

Lecture Notes on Coastal and Estuarine Studies

Managing Editors:
R. T. Barber, C. N. K. Mooers,
M. J. Bowman, B. Zeitzschel

Volume 6

Ecology of Barnegat Bay, New Jersey

Editors: **M. J. Kennish, R. A. Lutz**
1983. XIV, 396 figures. ISBN 3-540-90935-4

Contents: Introduction. - Physical Description of Barnegat Bay. - Nitrogen Distribution in New Coastal Bays. - Phytoplankton. - Macroflora. - Zooplankton. - Benthic Fauna. - Shellfish. - Shipworms. - Fouling Organisms. - Fishes of Barnegat Bay, New Jersey. - Commercial and Sport Fisheries. - Trophic Relationships. - Antropogenic Effects on Aquatic Communities. - Summary and Conclusions. - Appendix A: Bibliography of Unpublished Technical Papers, Theses, and Government Reports on Barnegat Bay, New Jersey. - Author Index. - Organism Index (Scientific Name). - Subject Index.

Volume 7
W. R. Edeson, J.-F. Pulvenis

The Legal Regime of Fisheries in the Caribbean Region

1983. X, 204 pages. ISBN 3-540-12698-8

Contents: Introduction. - The International Law Background. - National Legislation Relating to Fisheries. - Bilateral and Joint-Venture Fisheries Agreements. - Fisheries Administration. - Conclusions. - Tables. - Bibliography.

Volume 8

Marine Phytoplankton and Productivity

Proceedings of the invited lectures to a symposium organized within the 5th conference of the European Society for Comparative Physiology and Biochemistry - Taormina, Sicily, Italy, September 5–8, 1983

Editors: **O. Holm-Hansen, L. Bolis, R. Gilles**
1984. VII, 175 pages. ISBN 3-540-13333-X

Marine Phytoplankton and Productivity is a collection of authoritative articles designed to give the reader better insight into and understanding of primary production in natural waters and the cycling of organic carbon through heterotrophic microbial populations. Emphasis is placed on those environmental stresses which can affect the standing stock and metabolic activity of phytoplankton.

Volume 9

Osmoregulation in Estuarine and Marine Animals

Proceedings of the Invited Lectures to a Symposium organized within the 5th Conference of the European Society for Comparative Physiology and Biochemistry - Taormina, Sicily, Italy, September 5–8, 1983

Editors: **A. Pequeux, R. Gilles, L. Bolis**
1984. X, 221 pages. ISBN 3-540-13353-4

This book presents a comparative survey of the progress that has been made in understanding how the control of the body fluids is achieved in a wide range of estuarine and marine organisms. It centers essentially on the salt-transporting tissues - ranging from integuments and nephridia of molluscs and annelids to the salt gland of reptiles and birds - emphasizing the prominent part they play in the regulation of the body fluids' salt content.

MIX
Papier aus verantwortungsvollen Quellen
Paper from responsible sources
FSC® C105338
FSC
www.fsc.org

If you have any concerns about our products,
you can contact us on
ProductSafety@springernature.com

In case Publisher is established outside the EU,
the EU authorized representative is:
Springer Nature Customer Service Center GmbH
Europaplatz 3, 69115 Heidelberg, Germany

Printed by Libri Plureos GmbH
in Hamburg, Germany